RESIDENTIAL
INTEGRATION
SERIES

Residential Integrator's Customer Relations

TERRY TAYLOR, PhD
PRESIDENT, QUANTUM INTEGRATIONS

THOMSON
DELMAR LEARNING™

Australia Canada Mexico Singapore Spain United Kingdom United States

THOMSON
DELMAR LEARNING

Residential Integrator's Customer Relations
Terry Taylor, PhD

Vice President, Technology and Trades ABU:
David Garza

Director of Learning Solutions:
Sandy Clark

Executive Editor:
Stephen Helba

Product Manager:
Sharon Chambliss

Senior Editorial Assistant:
Dawn Daugherty

Marketing Director:
Deborah S. Yarnell

Channel Manager:
Dennis Williams

Marketing Coordinator:
Stacey Wiktorek

Curriculum Manager:
Elizabeth Sugg

Director of Production:
Patty Stephan

Production Manager:
Larry Main

Content Project Manager:
Nicole Stagg

Technology Project Manager:
Kevin Smith

Technology Project Specialist:
Linda Verde

Printed in the United States of America
1 2 3 4 5 XX 08 07 06

For more information contact Thomson Delmar Learning
Executive Woods
5 Maxwell Drive, PO Box 8007,
Clifton Park, NY 12065-8007

Or find us on the World Wide Web at
www.delmarlearning.com

Library of Congress Cataloging-in-Publication Data:

ISBN: 1-4180-1413-3

NOTICE TO THE READER

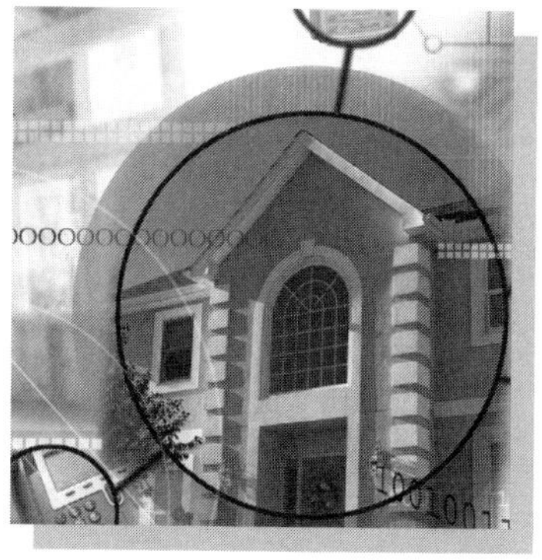

Contents

Series Preface

You've heard these catch phrases many times. Home theaters. High-Definition Television (HDTV). Distributed audio and digital video. Lighting and home control systems. Broadband Internet access and wireless networking. What do all of these terms mean? Simply stated, it's cutting-edge technology that is changing how we live—and will live—in our homes.

Why is this technology emerging? For one thing, the need for technical content and information is growing. But perhaps more importantly, people want the convenience and time- and cost-saving benefits of controlling their homes electronically as a way to spend more quality time with their families.

The residential integration industry is enjoying unprecedented growth in the installation of these new technologies. Due to the high demand, residential integration companies are seeking qualified individuals who understand the traditional fields such as new home construction, along with newer disciplines such as structured low-voltage wiring and data systems. Now is the time to join the bandwagon!

Recognizing the need for educational resources in this field, Thomson Delmar Learning is excited to present a four textbook suite, called the *Residential Integration Series,* which addresses this exploding industry. These four texts encompass many of the aspects of the residential integration industry to include the basics of the business and how to get started in it, customer service skills, project management, and finally a text on the information required to prepare for the various certifications that are currently offered within the industry.

More specifically, these textbooks are:

Residential Integrator's Basics is the foundational text for this industry. Here you'll find comprehensive information on computer networks, communications, home automation, cabling, wiring, and other topics. The final chapter addresses the Home of the Future.

Residential Integrator's Customer Relations pinpoints the types of customer service skills a residential integrator needs to be successful in this industry. These include working with both internal and external clients, working in teams (very important in this field), handling difficult client relationships, communications skills, training the client on both technology and equipment, and ensuring client satisfaction.

Residential Integrator's Project Management is based on *A Guide to the Project Management Body of Knowledge* (PMBOK), which is a widely accepted work published by the Project Management Institute (**www.pmi.org**). This text is divided into four sections, each covering a major phase of a residential integration project: The Foundation, Defining the Project, Planning the Project, and Executing, Monitoring/Controlling, and Closing the Project. This text is accompanied by a CD-ROM that contains project management templates specific to the industry that can be adapted to the needs of a particular company.

Residential Integrator's Certification provides in-depth coverage of the information required to prepare for the CompTIA HTI+ exam and the CEDIA Designer Classification Series Exam. This includes low-voltage cabling, high-voltage wiring, computer networking, audio/video, security systems, home controls, and other industry-related topics. Two appendices contain exam objectives for both exams.

The textbooks are pedagogically rich with chapter objectives, critical thinking questions, study tips, and other suggested activities, along with chapter summaries and review questions to help you learn and retain the material.

TextChoice
▸▸▸ THOMSON CUSTOM SOLUTIONS

Build Your Perfect Course Solution

It's your course, so why compromise? Now you can create a text that exactly matches your syllabus using **Thomson Custom Solutions** online book-building application, **TextChoice.** TextChoice allows you to easily browse and select content from leading Thomson textbooks and custom collections—even include your own content—to create a text that is tailor-fit to the way you teach. Visit TextChoice at **www.textchoice.com** and learn how Thomson Custom Solutions can help you teach your course, your way.

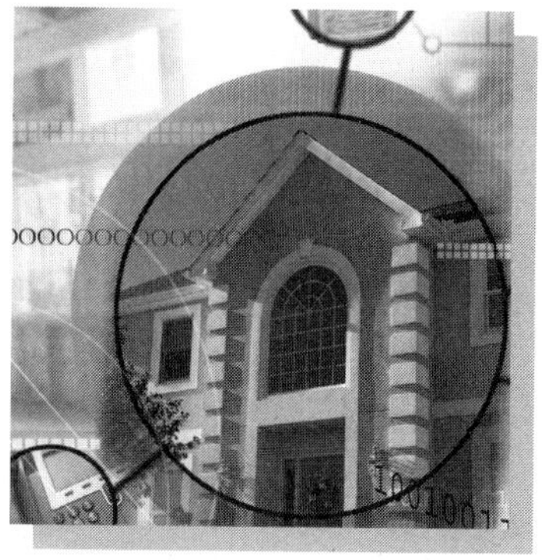

Preface

Residential Integrator's Customer Relations is part of a four-textbook series called *The Residential Integration Series*. The material in this textbook is intended for those in the residential integration industry who want to improve their customer relationship skills.

Poor customer relations, unfortunately, are a part of our everyday lives. We witness it more than we would like and often have little power to control the outcome of being treated poorly. In the residential integration industry, having employees who exhibit inadequate customer relation skills with the stakeholder (generally the homeowner) can spell disaster for an organization. Remember that oftentimes the residential integrator is the lone representative of the company at the customer site. He is a direct reflection of the company's mission, workmanship, attitude, and professionalism.

To mitigate potential problems, *Residential Integrator's Customer Relations* addresses the specific customer relationship skills that employees should have to be successful in this industry. This text provides in-depth coverage in the following areas.

Chapter Descriptions

Chapter 1, Introduction to Excellent Service, identifies the types of customers with whom residential integrators will interact, as well as defines excellent service within the context of the industry. It also discusses a company's mission, structure, and business strategy in broad terms and the reasons why it is important for all employees to embrace this information.

Chapter 2, Customer Connection Strategies, focuses on several key areas to include the attributes of an excellent service attitude, a critical component for personal and professional success. It also discusses personal strategies for connecting with customers, as well as understanding customer needs and expectations. The chapter closes with a dialogue about the importance of teamwork.

Chapter 3, Effective Communication for Excellent Service to Customers, defines the characteristics of clear communication in face-to-face interactions, on the telephone, in e-mail, and in written documents, all of which are crucial skills to master. Additionally, this chapter explains the characteristics of an effective listener and how to ask questions for understanding.

Chapter 4, Handling Difficult Customer Situations, addresses two important aspects of the customer-employee relationship—handling dissatisfied customers and handling the difficult customers. Troubleshooting and problem-solving skills are identified as ways to help diffuse volatile customer situations.

Chapter 5, Training the Customer, is of particular importance in the residential integration industry. Customers may know what they want in terms of home comfort, but they may not be technically savvy or understand the complexities of the equipment being installed in their home. This chapter outlines how to teach customers simple skills, and discusses key principles that can be applied in any training situation.

Chapter 6, Ensuring Customer Satisfaction, details effective ways to follow up with the customer, identifies sales techniques residential integrators can use with customers (very important in this industry), and discusses how to "go the extra mile" for the customers to ensure that the customer-company relationship remains strong after the installation has been completed.

Features

Chapter Objectives. Each chapter begins with a detailed list of the concepts to be accomplished within that chapter. This list provides the student with a quick reference to the chapter's contents and is a useful study aid.

Illustrations and Tables. Illustrations are provided to help the student visualize important concepts and procedures presented in the chapter.

Chapter Summaries. Each chapter is followed by a summary of the concepts presented in the chapter. These summaries provide a helpful recap and review of the chapter.

Key Terms. All key terms introduced in bold face type in the chapter are listed at the end of the chapter along with the definitions. This provides the student with an opportunity to check their understanding of all the terms presented.

Review Questions. End-of-chapter review questions reinforce the ideas introduced in each chapter. These questions ensure the student has mastered the concepts presented in the chapter.

Supplemental Material

The following supplemental materials are available with the text.

Faculty Guide with CD-ROM. The printed faculty guide offers a complete teaching package. Components for each chapter include, chapter objectives, key terms, classroom discussion questions, learning activities, teaching tips, chapter summaries, answers to the end-of-chapter review questions, and an outline of the PowerPoint® presentation.

CD-ROM. Each faculty guide has an accompanying CD-ROM that includes a PowerPoint® presentation, computerized test bank, sample syllabi, and suggestions for additional resources.

About the Author

Terry Taylor, PhD, President Quantum Integrations

Quantum Integrations, Inc. is a privately held company providing curriculum, instruction, and faculty development solutions for the education and training market. We are dedicated to assisting our customers in effectively providing the highest quality education to an increasingly diverse and challenging student/participant population. We do this through design of innovative new academic programs for colleges and universities, customized curriculum, faculty development courses and materials, and custom faculty development workshops. For more information about Quantum Integrations, please see our website at **www.qicurriculum.com.**

Introduction to Excellent Service

OBJECTIVES

After studying this chapter, you should be able to:

- Describe several different categories of customers commonly faced in the residential integration industry and describe common needs of individuals within each category.
- Define excellent service.
- Explain how service affects customer satisfaction, company success, and personal success as an employee of the company.
- Explain the importance of excellent service within the residential integration industry.
- Explain the purpose and importance of a company's mission or vision and organizational structure in providing excellent service to customers.

OUTLINE

Introduction

Think about the last time you had a service performed in your home. What was your experience? How did the service technician or installer make you feel? Did you feel like you were a priority? Did you feel appreciated? Did you feel respected? Did you feel as though you got your money's worth? Did the technician finish the project? Did he or she finish in the promised time frame? Did you feel as though you would refer your friends to this service provider? Why?

Now, think about the ideal situation. How do you want to be treated as a customer? What determines value of the product or related services you purchase? What factors determine whether you get your money's worth? The answers to these questions define either excellent or poor service to customers.

As a residential integrator, you will spend most of your time at a job site, which typically is or will soon be the customer's home. As such, you may be the only representative of your organization there on a daily basis. The customer will come to rely on you, not only for technical information, but for support in the project as well. Your ability to present yourself as a qualified member of your company's team is paramount. Your customer service skills need to shine, not only for you personally, but also for others you work with and represent in the entire project.

This text explores many areas of service to customers that will help you succeed as a professional and assist you in helping your company succeed with their customers. Providing excellent service to customers is not just a nice thing to do, it is a necessity in today's highly competitive environment and for highly sophisticated and demanding customers. How you treat your customers directly affects your company's bottom line. And, the bottom line means the difference between a healthy and ongoing company and one that is doomed for eventual failure. Excellent service is important. Customers pay attention to how they are treated and how you meet their expectations as an individual and as a company. You owe it to yourself, your company, and your customers to learn how to provide service with excellence.

Who Is the Customer?

Employees typically face many different types of customers on a daily basis. It is important to understand who your customers are, what they need, how you can provide them with excellent service, and how you can make them feel valued and appreciated. Let's explore the several different types or categories of customers you may face in your daily tasks. By understanding the different situations and needs of each customer category, you can provide your customers with the right kind of service and work to solve problems they commonly face—even before the problem occurs.

It is also important to recognize the cast of characters you must satisfy as the project evolves. Let's look at the installation of an integrated home entertainment system in a new home as the home is being built.

Initially, you must satisfy the customer (often the homeowner) plus a design team, which often includes the designer and architect on large projects and the builder on smaller projects. A slip at the project initiation stage can easily cost you and your company the job before you even get started. As the project progresses, the cast of characters shifts to include the various trade contractors and laborers who are involved on a daily basis. As the project nears completion, the focus shifts back to the builder and the customer. This is often the most stressful period for everyone involved. The builder is concerned about deadlines and ultimatums. The customer is exhausted from the entire process. The trade contractors are gone, and the installing dealer is left to finish the project. In this type of installation, most of the very expensive and sensitive electronics cannot be installed until the floors are sanded and varnished, which is the last step before the customer moves in. By this point, the customer intends to move in no matter what. The amount of work for the installer almost always exceeds the time between the floor completion and the customer move-in. This fact guarantees that the installers will be in the home with the customer as they are unpacking and settling in. This scenario represents just one workplace situation. As you can see, excellent service to the customer, the architect, the designer, the builder, and the trade contractors is essential for success.

The first step is to define these different customers in more detail. With this understanding, we can move on to techniques and strategies to best serve each group.

FIGURE 1–1 It is important to present yourself as a qualified member of your company's team. Customers pay attention to how they are treated and how you meet their expectations.

External Customers

An **external customer** is one who directly or indirectly pays for your products and services. A customer who directly pays for your products and services pays you with cash, check, credit card, or through a company account. In some cases, customers pay for your products and services indirectly through a third party. One example of an indirect customer includes taxpayers who pay for public education through their tax dollars. Another example is a homeowner who pays for your products and services through a contractor or builder. In some cases, you may never see the homeowner and will deal only with the contractor or builder. In other cases, you may deal with both the builder and the homeowner. The relationship with the customer is often intense, but typically short-lived, extending through to the completion of the project.

Potential Customers or Referrals

One subcategory of external customers is the potential customer, often called a referral. Businesses commonly spend large amounts of money recruiting new business. Marketing, advertising, and sales typically make up large percentages of a business's operating budget. The goal of these departments is to attract potential customers. In this industry, a very large part of the new business comes from referrals from satisfied customers or satisfied builders and contractors. Most professionals "put their best foot forward" with potential customers hoping that they will convert to new customers.

Potential customers include family and friends of customers and businesses already using your services. Word of mouth is a powerful tool if your customers are pleased with your products and services as well as with how they have been treated. Word of mouth can also be a powerful deterrent to potential customers. Remember that people typically tell more people when they are unhappy than they do when they are satisfied. Treating existing customers well increases the likelihood of them introducing potential customers to your business. Also, any person you meet who is not a customer is a potential customer. It is important to treat everyone as though they may soon purchase what your business is selling.

New Customers

A **new customer** is one who has recently purchased a product or service. They are the ones who are just now forming an opinion about your business. If the sales staff made a favorable impression and the initial functioning of the product, installation of the product, and overall customer service has been excellent, the stage is set for long-term customer loyalty and referrals to others. If any part of the initial sales process or the initial stage of the project has experienced problems, you could easily lose the entire job even before it begins. It is important to be aware when things are going unfavorably and to have strategies to help make things right quickly.

Existing Customers

An **existing customer** is one who has purchased your products and services and who resides on your customer list. These customers will look to your business first for additional products or upgrades and consistently refer others to your business. Remember that in this industry, the bulk of new projects comes from referrals from existing or former satisfied customers. These satisfied customers already know you exist. They understand what you do as a company. They have already purchased something from you. You do not need to spend money to find and recruit them. Compared to new customers, existing customers are very cost-effective. These customers should be treated like gold.

Exiting Customers

In every company, a certain percentage of the customers are walking out the door. An **exiting customer** is one who leaves for a definite reason, such as dissatisfaction or just in the natural course of life. These customers also deserve your attention and care. By fixing the problem that is making them leave, you may be able to convert them back to existing customer status. Even if you cannot fix their problem, by treating them with respect, they will often tell others who *will* become new customers.

Internal Customers

An **internal customer** can be an individual, a department, or a partner who needs something from you and on whom you rely to serve external customers successfully. In most projects in this industry, external

FIGURE 1–2 It is important to satisfy internal customers, such as builders and architects, in order to ensure the long-term success of your company.

customers include installers and technicians waiting for design, instructions, and parts. Additionally, the designer, architect, builder, trade contractor, and other laborers could be internal customers. Just as it is important to satisfy external customers in order to gain and keep their business, it is important to satisfy internal customers in order to ensure the long-term success of your company.

What Is Important to Customers?

Each customer has a need that they want filled. Every player in a project has a place in successfully filling the need. The customer's need may be to feel good or to make someone else feel good, or the need may be to solve some problem the customer has. Let's explore some of the possible needs that your internal customers and external customers may have. The more you understand about your customers, the better you will be able to provide them with excellent service.

Needs of Internal Customers

Your internal customers (those who rely on you to help them do their job or who receive products, services, or information from you) have very definite needs that you can help them achieve.

- They need to have competent and efficient people doing quality work so that the entire project is completed successfully.
- They need to look good in the eyes of their supervisor, their coworkers, and the external customer.

- They need individuals who are easy to work with so that their day is less stressful with fewer hassles.
- They need to be surrounded by individuals with great attitudes and enthusiasm so that their day is enjoyable.
- They need to work with people who are good communicators so that miscommunication and misinformation is avoided.
- They need to work with people who are reliable, trustworthy, punctual, and honest so that they do not have to worry about the many problems caused by unreliability and dishonesty.
- They need to feel appreciated and valued.
- They need to work with individuals who ultimately provide excellent service to the external paying customer so the external customer will remain a loyal customer, pay their bills, return for additional products and services, and tell others how great their entire experience was.

Needs of External Customers

Your external customers (those who pay you directly or indirectly for your products, services, or information) also have very definite needs that you can help them achieve.

- They need individuals who provide high quality work in an efficient manner.
- They need to feel that they have received more than what they paid for and that there is value to what they have purchased.
- They need to feel appreciated and valued as a customer.
- They need attention.
- They need to feel that the individuals they are paying are confident and enthusiastic about their jobs and exhibit great attitudes as they perform their tasks.
- They need to be served by individuals who are reliable, trustworthy, honest, and punctual because their time and money are important to them.
- They need to be served by individuals who are good communicators so that miscommunication and misinformation are avoided.

Why Do Customers Leave?

It is important to understand why customers leave a specific company. There are only four possibilities according to author Hughes of Database Marketing Institute (Hughes, 2005).

1. They die or are no longer buying what you sell.
2. They are unhappy with the price.
3. They are unhappy with the product.
4. They are unhappy with how they were treated.

Unless you work in management, marketing, sales, advertising, research and development, or production, you probably have little say regarding price or product. However, as a first-line representative of the company with the customer, you have great influence on how customers are treated. Assuming that the customer is buying what you sell and is satisfied with the product and your work, you may be the key employee in keeping the customer satisfied.

Critical Thinking Question

Think about your purchasing habits and experiences.

- To which businesses are you a loyal customer?
- Why do you stay loyal to that company?
- How are you treated?
- When have you left a business?
- Why did you leave?
- How were you treated?

List customer service "do's" and "don'ts" that you have learned just from your own personal experiences.

Study/Career Tips

Take a moment to think about the different kinds of customers you may face personally in the course of your job. List all the different categories of customers and identify their specific needs as they relate to your industry. For each need, write down ways that you can help them meet their needs. Also for each customer, identify how providing each type of customer with excellent service will help your business and you personally to succeed. Be as specific as possible. We will address this question again later in the text.

Excellent Service Defined

Excellent service is the total effect of the products, services, and information you provide to your customer combined with how you treat the customer before, during, and after the sale or service. Excellent service is an attitude and should become a habit.

Important: *The customer is the reason you have a job.* With today's intense competition in most industries, customers can easily take their business elsewhere if they are not satisfied.

Before moving on, let's explore a few basic premises and assumptions related to customers.

- Customers may pay less attention to the product or service they are purchasing and more attention to the smoothness of the process and how they are treated.
- Price is rarely the most important issue. Customers will pay more for exceptional service. Research suggests that they will pay as much as 30 percent more.

- Installation companies usually have one opportunity to perfect a transaction with a customer, much like a builder or other contractor. In many installer companies and other areas of the residential integration industry, the opportunity for continued business after the completion of the project is limited. However, there is a lot at stake during the primary transaction.
- The satisfied customer is a highly effective marketing tool for your company. This customer marketing and referral arena is where the continued satisfaction of a customer is essential. Few companies can survive without a growing referral base.
- It is the responsibility of each company employee to treat customers well.

Service Goals for the Company

The goal of any successful business is the bottom line—to make a profit. If a business is not making more money than it costs to produce or sell the products or supply the services, it will fail. The business must sustain this positive revenue stream over the long term as well. In the big picture, this means that a business must either attract new customers or sell existing customers more products and services. In this industry, the most profitable option is to attract new customers through referrals; therefore, the goal is a satisfied customer who will provide you with references and who will actively function as a referral agent. Additionally, well-run projects are always more profitable; therefore, an important goal is to ensure that projects run smoothly without problems and with great customer satisfaction.

Think about how important this really is over the long term.

- Typically, individuals increase their salaries throughout their career; therefore, they can afford more and higher priced products and services.
- As customers gain knowledge about technology products, they typically want to buy more advanced (and higher priced) products. With technology, there will always be something new and better to sell to them.
- Families also grow, requiring more and more varied products and services. Kids seem to want the latest technology.
- Satisfied customers tell others. Marketing is expensive. Simply keeping your existing customers happy is almost free marketing.
- Dissatisfied customers tell even more people. The costs and challenges of marketing increase if the business must overcome a negative reputation.

Customer Service Goals for the Employee

Good **customer service** is definitely a plus for the customer and a necessity for the company. It is also important for individual employees.

- A culture of excellent customer service typically improves the environment in which you work. This results in more job satisfaction, more fun in the workday, and greater enjoyment.
- Individuals who practice excellent customer service often see that there is less stress in their daily job. Internal and external customers are nicer and easier to deal with. Common hassles and frustrations are minimized. In general, things just tend to run more smoothly.
- Individuals who provide excellent customer service are typically noticed. When employees are noticed for positive reasons, they tend to receive better workplace rewards such as increased salaries, fast promotions, employee recognition, and better bonuses.

Critical Thinking Questions

Think about your current job or a job you have had in the past.

- What things typically caused you to have a bad day? List as many factors as you can on a piece of paper.
- Now, how many of those factors could have been avoided if you or someone else had practiced excellent customer service?

Customer Service Goals for the Customer

Put yourself in the place of the customer for just a minute. Though the ultimate goal is to get the customer to eventually buy more from the business and to tell their friends or colleagues to buy from your business, the customer does receive real benefits from excellent customer service. Their goal is a pleasant experience concluding with the successful installation of a complete system as they expected.

- They get their needs met in terms of buying excellent products, services, or information.
- Especially in technology areas, the customer can gain valuable information that they can use to solve their technology problems in the future.
- A culture of excellent customer service reduces the common hassles and frustrations faced in many business transactions. A business that provides great customer service can actually make the business transaction a pleasant experience.
- Employees who provide services with great attitudes and enthusiasm usually improve the customer's day.

The Customer Service Attitude

As stated before, it is the responsibility of each employee to treat the customer well and to ensure that the customer is satisfied. Treating the customer well starts with the very first impression, whether it be on the phone, in a face-to-face encounter, or in an e-mail or other form of written communication. Customers can sense the commitment to customer satisfaction from the very first moment. If the first impression is favorable, the stage is set for excellence. If the first impression is not favorable, the customer will tell others quickly.

Each employee's attitude represents the customer service commitment of the entire company. Even one employee with a poor attitude puts a damper on the entire organization. An excellent customer service attitude is established by appearance, words, nonverbal behaviors, and actions. Attitudes are long lasting.

Study/Career Tips

As you go through the chapters of this text, reflect on your own customer service skills, attitudes, and areas for improvement. Remember that your goal is to create excellent customer service habits and develop the skills required to meet the many needs of your customers. You will benefit both personally and professionally if you do so. Consider acquiring a notebook and taking detailed notes as you answer the following questions on each customer service concept we discuss.

- In what aspects of this skill do I currently do well? How can I use this existing skill to my advantage in the workplace?
- In what aspects of this skill do I need some work? What specifically do I need to improve? What steps can I take for improvement?

Write down your responses and follow your action steps. Then, review your progress and write down how you have improved and ways you can continue your improvement.

Understanding Your Company's Mission and Structure

To be equipped to provide excellent customer service, it is important to fully understand your company. Two important elements that help to guide how you operate within your company are the company's mission statement and the company's structure. Let's explore how these elements factor into providing excellent customer service.

Company Mission Statement

No doubt your company has developed a company mission statement that identifies the purpose of the business and guides how business is conducted. A **mission statement** is a concise declaration of the kind of business you are in and it identifies the customers you serve. A mission statement is sometimes called a company vision. Typically, there are both an internal and an external dimension to the statement. The internal dimension describes the actual products your company provides and the services your company performs in support of these products.

The external dimension describes the specific market(s) you serve by identifying the customers and why they would want to buy from your company or use your company's services. The main purpose of a mission statement is to focus both employees and customers on the main goals of the company. Typically, mission statements are published internally to all employees and externally on marketing materials such as brochures and Web sites. Marketing managers, sales staff, business managers, production managers, and so forth can use the mission statement to appropriately designate their limited resources to best meet the company's goals.

A mission statement answers many of the following questions:

- What is the purpose of our business?
- Who are the main users of our products and services?
- What are our principle market segments?
- What differentiates our company, products, or services when compared to those of our competitors?
- What philosophical issues are important to us?
- What concerns do we have as a company and how do we address these concerns?
- How do we build bridges between our company and our customers?
- Where is the company headed and how are we going to get there?
- What kind of business opportunity should the company say "no" to?

Though a company's mission may evolve over time as the economy and technology change, it is basically independent of time and immediate market conditions. As an excellent employee, you should know the company's mission statement and constantly think about how to apply the mission to the daily job.

Critical Thinking Questions

Go to the Internet and search for a technical company's mission statement. Try using the words "mission statement" as the search term. Or, go to the Web site of a technical company you are familiar with and search the site for their mission statement. Consider choosing companies you would soon like to work for.

- What questions does the mission statement answer?
- How does the mission statement address the customer?

Compare and contrast the mission statements of at least four different companies.

- What trends do you see?
- How are the company's statements different?
- If you were going to develop your own company's mission statement, what elements would you include? Exclude? (Based upon the examples you researched.)

Company Strategy and Structure

A goal of any company is to maximize profits. To do this, the company must have a specific plan for doing business. This plan is called the **business strategy.** The company must also have division and specialization of labor. A business strategy changes as competitors change, as markets change, as technologies change, and as regulations on a company change. Basically, the business strategy describes how a company is going to do business within the current economy and market.

The activities of each department and individual must be coordinated and managed. A company's **organizational structure** describes the various tasks and positions within a company and how these positions interrelate with each other.

Ideally, a company's organizational structure should be designed to accomplish the following objectives:

- To provide for effective managerial leadership within and between departments and from the highest position, such as the CEO or President, to the lowest positions
- To allow each employee to know exactly what they are supposed to do, how they are supposed to do it, and how it fits into the big picture
- To allow all employees to successfully carry out the company's mission
- To define how all employees interact with each other and the customer
- To define the lines of communication, chains of command, and reporting relationships
- To define who is responsible for each required task in the company and assure that no task falls through the cracks
- To foster teamwork within the company for the best service to the customer
- To help management analyze the effectiveness and efficiency of operations, systems, policies, procedures, and regulations.

The larger a company is, the more departmentalized the company becomes. A department is a group of employees who work together to accomplish a major function of the company. Examples of departments include personnel or human resources, production, research and development, sales, marketing, advertising, finance, distribution, purchasing, service, and so forth. The ideal company structure organizes these departments so that they function and communicate in the most effective and efficient way. In this industry, most businesses minimally have the following departments:

- **Sales and Marketing:** The Sales Department is responsible for selling the services to the customers.
- **Design and Engineering:** Those employees involved in design and engineering are responsible for the technical aspects of the equipment, installation, and maintenance services.

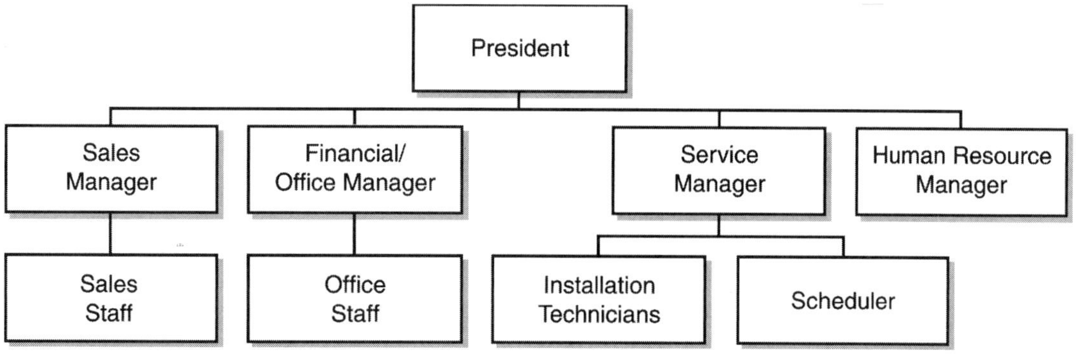

FIGURE 1–3 Organizational Chart

- **Accounting/Administration:** Those working in the office are responsible for accounts receivable, accounts payable, personnel tasks such as hiring and firing staff, vendor relations, compliance with regulations and laws, accounting, and tax planning.
- **Operations:** Operations tasks include installation, project management, service, material acquisition and inventory, vehicle maintenance, tool maintenance and inventory, and so forth.

A primary company structure tool is the organizational chart. An **organizational chart** is a diagram that lists all departments and positions within a company and illustrates how these positions interrelate and communicate with each other. Detailed organizational charts include actual names of employees filling each position. See Figure 1–3 for a sample organizational chart.

In order to provide excellent service to all of your customers, both internal and external, you must understand how your company operates and how it is organized. A look at an organizational chart provides an outline of how your company is aligned and how work flows from one department to another. With this understanding you can provide your customers with accurate information, step in when needed to facilitate problem solving, explain company processes and procedures to minimize frustration, and represent your entire company competently. Understanding your company's structure also allows you to see the various lines of communication and relationships so that you can work within the structure efficiently and effectively.

Critical Thinking Question

Go to **www.google.com** and click on the "Image" tab. Use the key words "organizational chart" to view many different ways that companies are organized. Study several of these organizational charts. Which of the following types of organizational structure do you think is more effective? A *tall structure* where there are more levels of management or a *flat structure* where there are fewer levels of management? Explain your answer.

Unlike government agencies or large corporations that seem to erect barriers to all but the most persistent customers, the residential integration industry is based entirely on the personal relationships and good will of customers. The nature of this industry does not allow for clever marketing to replace sincere concern. This industry consists of thousands of small businesses constantly establishing personal relationships with valued customers who often hold the dealer's company in their hands. It is not unusual to observe the future of the company linked to the successful completion of the installation of a specific product. Simply delivering what is in the contract is not sufficient.

Customer service is not just a good idea; it is the law if a business wants to survive. A company in this industry is often small and the link between cause and effect is almost instantaneous. An unhappy customer will not pay. The nature of a transaction in this industry is one of a handful of large transactions. This fact makes the company itself vulnerable. If a large customer refuses to pay in a timely manner, the employee may not have a business to report to. A mediocre result may get the final check, but it will not get the enthusiastic referrals required to propel the company forward.

Summary

In this chapter, we discussed some foundational aspects of customer service in order to set the stage for more detail in the next chapters. First we identified the different categories of customers and the unique needs of the different individuals you may face on a daily basis. We looked at specific reasons that customers leave companies and go to the competition, and we identified some basic assumptions about customers for any business. We also reviewed many reasons that excellent customer service is important and generally defined an excellent customer service attitude. Finally, we briefly discussed some business concepts that companies use to address service to customers: a company's mission statement, the business strategy, and the organizational structure.

Important points in this chapter are:

- It is essential to understand the different types of customers you might face on a daily basis: external customers, new customers, existing customers, exiting customers, potential customers, internal customers, and regulated customers.
- Each customer has specific needs that must be addressed to their satisfaction in order for them to remain loyal customers to the business.
- Customers typically leave a company for specific reasons.
- Customer service is the total effect of the products, services, and information you provide your customer combined with how you treat the customer before, during, and after the sale.
- Providing excellent customer services has benefits for the customer, your company, and you personally.
- Businesses often explain who their customers are and how they want to treat customers in their mission statements.
- A company's business strategy and organizational structure affects the kind of customer services that is provided.
- In order to serve customers with excellence, all employees should thoroughly understand the company's mission or vision, the business strategy, and the organizational structure.

Key Terms

Business strategy A business strategy is a company's specific plan for doing business.

Customer service Customer service is the total effect of the products, services, and information you provide your customer combined with how you treat the customer before, during, and after the sale.

Excellent service The total effect of the products, services, and information you provide to your customer combined with how you treat the customer before, during, and after the sale or service.

Existing customer An existing customer is one who has purchased your products and services and who resides on your customer list.

Exiting customer An exiting customer is one who leaves for a definite reason, such as dissatisfaction, or just in the natural course of life.

External customer One who directly or indirectly pays for your products and services.

Internal customer An internal customer can be an individual, a department, or a partner who needs something from you and on whom you rely to serve external customers successfully.

Mission statement A mission statement is a concise statement declaring the kind of business you are in and identifying the customers you serve. It is sometimes called a company vision.

New customer A new customer is one who has recently purchased a product or service.

Organizational chart An organizational chart is a diagram that lists all departments and positions within a company and illustrates how these positions interrelate and communicate with each other.

Organizational structure An organizational company structure describes a company's various tasks and the positions within a company and how these positions interrelate with each other.

Review Questions

1. Identify at least seven different categories of customers who might be faced in this industry on a given day. What are the needs of individuals within each category?
2. Why is it important to consider individuals within your own company as customers?
3. What benefit is there to treating regular customers with excellent customer service?
4. How does excellent or poor customer service affect potential customers?
5. What are at least five reasons customers leave to purchase products and service from the competition?
6. How does excellent customer service affect you personally as an employee?
7. Why is it important to be aware of your company's mission statement?
8. Why is it important to understand thoroughly your organization's structure and organizational chart in order to provide excellent customer service?

Customer Connection Strategies

After studying this chapter, you should be able to:

OBJECTIVES

- List the attributes of great service.
- Identify personal strategies for connecting with customers.
- Explain the importance of understanding customer needs and expectations.
- Explain the importance and implications of excellent service to internal customers.
- Identify traits of a project team player.
- Explain how effective teamwork with other teams leads to effective service.

OUTLINE

Introduction

Attributes of an Excellent Service Attitude

Personal Strategies for Connecting with Customers

Excellent Teamwork

Understanding Customer Needs and Expectations

Introduction

Great service starts with a great attitude about your job, yourself, and your customer. The purpose of this chapter is to highlight some important characteristics that lead to an excellent service attitude. Excellent services allow employees to connect with their customers and to meet their needs in the most pleasant and efficient way possible. This chapter also discusses the importance of treating work projects as team projects and explores the many characteristics required to be an effective project team member.

Attributes of an Excellent Service Attitude

Attitude really is *everything* when it comes to dealing with customers. Attitude sets the stage for all other actions. And, a positive and service-oriented attitude can overcome many problems, dissatisfactions, or mistakes. It can also help in building long-term and loyal relationships with coworkers, collaborators, and customers.

Attitude is also a key factor in first impressions. It tells how you feel about yourself, how you feel about your job, and how you feel about the customer. It reveals confidence in doing an excellent job or fixing a problem, or it reflects a sense of incompetence.

Finally, attitude is contagious, not only to other employees or contractors with whom you work, but also to the customers themselves. Remember that customers can have their own problems (including attitude problems) and you have an opportunity to improve the customer's day by your positive outlook.

Critical Thinking Questions

Think about your own attitude in your school or job right now.

- Is your attitude positive?
- Does it reflect a sense of confidence in getting things done quickly and competently?
- Does it reveal unhappiness or disinterest or insecurity?
- If you were a company owner, would you want your current attitude to represent your company with existing or potential customers? Why or why not?

Attitude is made up of several elements: confidence, competence, appreciation, empathy, honesty, reliability, responsiveness, patience, open mindedness, and courtesy. If presented properly, these elements result in the assurance that you are capable of doing what you say you can do. Let's look at each of these elements independently.

Confidence

Do you project confidence that you can do your job, solve a problem, or find the information necessary to achieve the customer's goals? What do the people you work with think? Do they have confidence that you can do the job?

Conveying confidence is important for both your internal customers (contractors, coworkers, subordinates, supervisors, etc.) and your external customers (those paying for your services either directly or indirectly).

Your internal customers need to have faith that you know what you are doing. Their jobs and reputations are on the line along with your own. Your external paying customers are shelling out significant resources to have a task completed. They also need to feel that you have the skills and know-how to make their sacrifice of money, time, and inconvenience worth it.

A high level of confidence lowers stress and anxiety in those around you. Lowered stress and anxiety leads to calm and rational thinking, better relationships, and improved patience, which are all key ingredients to a pleasant and productive work environment for everyone involved.

Confidence is fostered by several behaviors:

- Know how to do your job well.
- Know that the products and services you provide and represent will meet the customer's needs and expectations.
- Define and understand the project parameters in advance.
- If you do not know how to do something or do not have sufficient product knowledge, learn it and practice your skills often.
- Believe in your own abilities to complete the task, solve the problem, or find needed information.
- Control your self-talk so that you treat yourself (and others) in a positive way.

Critical Thinking Questions

Think of a skill in which you have a high level of confidence. Now think of a skill in which you are very insecure.

- How do you behave differently in each skill?
- How would an observer be able to tell whether you have a high or low confidence level in each skill?

Be very specific.

Competence

Are you skilled in the tasks you are asked to do? Do you have the appropriate knowledge? Are you efficient so that you can do the tasks with ease and quickly? If there are skills in which you need improvement, do you know where to go for help? How will you improve your competence throughout your career? If you were asked to develop new skills, how would you go about doing this?

You must know how to perform the tasks of your job well. You must understand fully the products and services you represent. You must keep current with technology. You must know where to find information quickly and how to effectively communicate that information to others. Being highly competent tends to increase confidence. The more you know about what you are doing, the better service you can offer your customers.

FIGURE 2–1 It is important to have a thorough understanding of the product and services you represent.

Developing a high level of competence is not always easy. Often it requires additional study on your own time and effort beyond your day-to-day tasks. And, developing competence never stops. This fact is reflected in the need for life-long learning. In this context, **life-long learning** is the idea that a professional must continue developing knowledge and skills throughout their life in order to maintain competency for the workplace. Technologies constantly change, the base of knowledge surrounding your industry continuously grows, and your company is no doubt continuing to add new products and services to their list. It is your job to take the steps needed to maintain a high level of competence in your job. It may not be written down in your job description, but it is your responsibility nonetheless.

Consider the following activities to continue to learn throughout your career:

- Take continuing education courses, trade shows, or seminars related to your job.
- Become a member of your industry's many professional organizations.
- Achieve the appropriate credentials related to your position.
- Actively seek out the knowledge and skills of more experienced workers around you.
- Learn to use the Internet effectively to find information related to your industry.
- Keep current with your company's products and services by reviewing catalogs, Web sites, product information, manuals, and so on.

Critical Thinking Questions

- In what areas of your industry do you think you will need to continually improve your knowledge and skills?
- Where will you find the information?
- How will you continue to build your competence?

Be very specific about the available resources and the steps you could take. Do some research to find more information.

Appreciation

Do you appreciate your customers (both internal and external)? How do you show your appreciation to others? Do you appreciate the skills and know-how you have developed yourself?

Appreciation is a mindset and is established through self-talk. **Self-talk** is what you say silently to yourself as you go through the day or are faced with difficult situations. Self-talk includes all the things you are saying to yourself as you work with a customer and complete your tasks. What you are saying to yourself always comes out in some type of behavior to your customers. It might be in a tone of voice, in an action, or in an easily perceived attitude. You choose what you think about your customer.

> Negative or condescending self-talk: *She has no clue about all the things this system can do! What an idiot.*
>
> Positive self-talk: *This is a great opportunity to help her learn about all the things this system can do! Part of what she bought was the training; she is paying for it every bit as much as she paid for the TV. We owe it to her. She is going to be so glad she bought it.*

The kind of self-talk you choose is just that—your choice. Positive self-talk sets you up for a positive attitude and will be followed by positive words and actions. Negative self-talk does just the opposite. It sets you up for a negative attitude followed by damaging words and ineffective actions.

Appreciation also includes how you feel about yourself and the efforts you have made to become successful. Self-talk applies here, too. With negative self-talk you set yourself up for failure. Positive self-talk can provide just the right amount of confidence and motivation to succeed.

> Negative self-talk: *I have no clue how to fix this problem. I might as well give up now before I waste any more time.*
>
> Positive self-talk: *I haven't seen this problem before. This will be a good opportunity to test my skills. Let's see what I can find out.*

Once you make a habit of believing in yourself and your abilities, words and actions will follow accordingly. You will see that you actually find ways to succeed.

Study/Career Tips

Notice your self-talk as you go through your day.

- What are you saying to yourself?
- What are the resulting words and actions?
- How does your self-talk affect your attitude?

If you find that you have a lot of negative self-talk, consciously change it to something positive. In a short time, you will find that you will not have to think about it. Positive self-talk will become a habit.

Empathy

Do you listen to your internal and external customers with empathy and truly want to help to meet their needs? How do you show empathy to those around you? What are the effects when you show empathy?

Empathy does not mean agreeing with everything your customer says. Nor does it mean promising to do everything your customer wants. Empathy does mean understanding what your customer needs and where they are coming from. **Empathy** is putting yourself in the other person's shoes, so to speak. Whether you agree with them or not, the customer does have a right to their opinions and point of view. They also have a right to be heard. Empathy is realizing this and acting and communicating accordingly.

The first step in empathy is to listen carefully to the customer without forming an opinion or making a judgment based on your own point of view. Taking turns to speak is not really listening. It is just being polite. Really listening means that you do whatever you need to understand the customer's concerns, problems, questions, and point of view. Really listening to a customer does two important things: (1) it gives you the information you need to solve the problem or meet the need, and (2) it affirms in the customer that they have been heard. This affirmation is critical and will go a long way toward building their trust in you. Once the customer trusts you, you then have the opportunity to address the issue, and/or complete your job successfully. You may be able to provide more information to help the customer fix the problem. You may change the customer's viewpoint by showing that they really didn't want what they initially thought was best. You may simplify what was initially perceived as a complicated issue. Basically, you must listen to the other person before you can expect them to listen to you.

Empathy is important in everyday dealings with customers, but it is especially important when the customer is frustrated, angry, or upset. When people are upset, their ability to think clearly or logically is often diminished. They often say irrational things, demand unreasonable responses, or behave inappropriately. Empathy diffuses emotion. When an angry customer experiences empathy from an employee, the typical response is to calm down and to return to rational and reasonable thinking. With rational and reasonable thinking, most problems can be solved satisfactorily.

Some common phrases that show empathy include:

Wow! I would be mad too if the product broke in the first week.

This problem is definitely an inconvenience. Let's see how to fix this as quickly as possible.

This static is really frustrating. I can understand how you feel.

I know it is frustrating when there is this much of a delay. Let me explain why our process is so important.

Be specific in your statements and try to avoid being vague. Identify the emotion that you think the customer is feeling. Try to also identify the source of the emotion.

A word of warning: If you try to show empathy without sincerity, you only make matters worse. Your customers are not stupid. They will see through your insincerity and feel patronized.

Study/Career Tips

As you go through your day, consciously try to think about what others around you are feeling. Think about how you might respond differently in the situation by thinking more how the other person feels. Develop a habit of empathy.

Critical Thinking Questions

- How might a lack of empathy affect your relationship with your customers?
- How might this affect your ability to provide excellent service?

Be specific.

Honesty

Are you honest with your customers and do your words and actions convey honesty? Do you strive to build trusting relationships with your customers? Specifically, how do you try to be honest with your customers? What are the consequences of dishonesty?

Honesty is an essential element in business success whether you are the CEO or on the lowest rung of the ladder. And, trusting relationships with customers foster customer loyalty and ongoing business.

Customers need to trust that you will be honest with them. They need to trust that you are providing them with the right information or are selling them the right products and services to meet their needs. Dishonesty is pretty easy to detect. It comes through clearly in words, tone, and body language, not to mention in customer eventual dissatisfaction when the dishonesty is found out—and dishonesty is *always* found out sooner or later.

To be honest, you must say the truth, follow through on promises, and state that you do not know the answer when appropriate. Customers will quickly determine that you (and the company you represent) are trustworthy and will continue to give you their business, often even when you cannot fix a problem or meet the immediate need. And they will tell others about how they were treated.

Employees get into "honesty" problems in many different ways. The following list reflects only a few examples that you might have experienced yourself:

- Promising that a product will meet a need when you know it won't
- Underestimating a wait time for service when you know it will be longer
- Stating that a product will be available when you know delivery has been delayed
- Underestimating the cost of a repair when you know that it will cost more.

There are some basic tips that will help to ensure that the customer sees you as trustworthy:

- **Over estimate wait times.** If you serve the customer faster than expected, the customer will be thrilled.
- **Never promise anything you are not sure you can deliver.** If necessary, tell the customer you need to check your facts with others and get back to them (and then get back with them when you say you will).
- **Don't tell a customer what they want to hear unless it is the truth.** Instead, tell them the facts and work on solving any problems.
- **Refer to someone else who can help, if necessary.** If your product or service cannot meet the customer's immediate needs, tell them so and refer them to a different vendor who can meet their needs. Though you may lose the immediate business, you may gain future business with your honesty.
- **Explain your actions thoroughly and in terms your customer can understand.** Sometimes a customer may distrust you simply because he or she does not have enough information to know that you are doing what you are supposed to be doing or giving accurate information.
- **Present yourself as working on behalf of the customer.** For example, address the customer's needs and offer information or make recommendations related to those needs.
- **Provide the advantages and disadvantages.** Tell the advantages and disadvantages of an product or service option so that the customer can make informed decisions.

This industry has a unique capacity for being misunderstood. Technology is a curse when it comes to expectations. There is almost always a conflict between what the customer thought or might have been told they were getting and the ultimate result. It is important to try to manage the customer's expectations by under promising and using strategies to ensure that the customer understands what is being purchased, what the equipment or system will do, and what it will not do.

What are five additional "honesty" problems that you might experience in your field? Add to the list based on your own observations and experiences.

Critical Thinking Question

Reliability

How do you rate yourself in terms of reliability? Do you *always* do what you say you will do? Can customers depend on you and your products and services to meet their needs? Are you regularly late, or do you typically run on time? How important is being punctual to you?

Reliability is essential to both your internal and external customers. Customers want products and services they can depend on. And they want to be able to depend on you too. Basically, reliability means doing what you say you will do when you will do it. If you have scheduled a service call for 9 AM, then show up at 9 AM (not before, unless you call to ensure it is convenient, and not after.) Obviously, things happen that are out of your control. In these cases, it is important to notify customers accordingly. It is also important to do everything in your power to control the situations. Good planning, accurate estimating of a project's time needs, accurate estimating of travel time, and organization all help in this control.

Reliability is also reflected in availability, prompt replies, quick follow-up, and fast work. Reliability implies accountability for actions and any potential problems. In order to be accountable, you must be available. Can customers contact you? Can your coworkers or office contact you easily? If a message must be left, do you respond to it quickly?

Keys to reliability are listed here. Add your own keys as they relate to your specific job.

- Ensure that customers and coworkers can contact you.
- If a message must be left, check your messages often and respond immediately.
- Respond to e-mail and voicemail immediately.
- If you are out of touch for a specific time, let people know so they won't be disappointed by your lack of response.
- Carry a calendar, notebook, or PDA (personal digital assistant) to keep track of your appointments.
- Use an alarm to remind you of important meetings (some watches and cell phones have this feature.)
- Show up on time—not before or after the agreed upon time, unless it has been OK'd ahead of time.
- Learn to accurately estimate how long it will take to complete a task, receive a part, schedule an appointment, and so on. Don't guess. Wait to give a time estimate until you have all the information.
- Learn how to do the job right to begin with. Reliability also means being able to depend on the quality of work.

Critical Thinking Questions

- What are specific problems an employee might cause for internal customers by being unreliable?
- For the company as a whole?
- For the customer?

Responsiveness

Do you respond to your customers quickly, accurately, and with the goal of meeting their needs and answering their questions? How do you show your customers that you are responsive? How does responsiveness impact your ability to provide excellent service?

There is nothing more frustrating than being passed from one person or department to another without ever getting what you need. Yet, this is often the experience of customers in many businesses. There is either an endless maze of numbers to select from on the phone or the customer is continually transferred from one person to another (translated as "put off as someone else's problem"). If this is the case with your own business, what can you do personally? How can your customers reach you directly? How can you make the seemingly endless maze more manageable? Is there a specific extension to call? Your responsiveness may not always be in your full control, especially if others screen your calls or messages. If this is the case, work closely with those who have control of your communications to ensure that you receive messages and calls immediately. Never keep a customer waiting longer than necessary. For good service, responsiveness is at the top of the list of key elements.

Responsiveness not only means that customers can find the people they need to solve a problem, it also means that these people get back with customers quickly with the right information that solves their problem. It also assumes a willingness to take care of the customer or intervene on their behalf. Responding with the easiest answer does not always mean great service. For example, if a customer asks you if you have a specific product and you really don't know if your company has that product or not, the easiest answer is to say no. Though you responded to the customer's question quickly, you have not met the customer's need to find the product. A simple phone call may find that a different department of your company does actually sell the product. Responding quickly is not necessarily responsiveness. Every customer problem or question is an opportunity to provide excellent service and to foster an even deeper customer loyalty.

There are many tips for being responsive to customers. Some of them are listed as follows:

- Make it a habit to answer your phone rather than let the call go to voicemail. However, do not answer a call if you are engaged in a conversation with the customer.
- Make it a personal policy to respond to your customer's phone calls or e-mails within an hour or so.

- Try to anticipate your customer's questions and problems so that you can begin working on them quickly, even before you return your customer's call or e-mail.
- Complete your tasks as quickly and efficiently as possible.

Patience

Are you patient with your customers who are obviously not as skilled and knowledgeable as you in your technical area? How do you demonstrate your patience? In what situations do you become impatient? When this industry was forming, the customers were "early adopters" of the technology and often as familiar with the systems as the installers. However, now that the market for technology has moved into the "mainstream" of consumers, customers are not experts. They are paying you to be the expert. Unfortunately, many installers are still treating their customers as though they were the highly knowledgeable early adopters of a decade ago when, in reality, many customers actually know (or care) very little about the technology. They really just want the system to do what it is supposed to do and make their life either more convenient or more entertaining.

Patience is a virtue, so the saying goes. Patience means to endure without complaint and with calmness, self-control, and tolerance. In many customer situations, being patient is not easy. Customers often do not know what they are talking about or cannot communicate their question or need. In many instances, customers do not know the first thing about operating the equipment they have purchased. The common phrase says: There are no stupid questions. Of course, there are stupid questions; and if you work in your industry long enough, you will no doubt hear plenty of them. The key to excellent service is to not make the customer feel stupid. Patience is essential. Nobody likes to feel stupid and when an employee shows impatience, emotions rise and frustration sets in.

Impatience probably means that you have some expectations about the customer. Add to the following list as you see the issues and expectations arise in your own job:

- They should do something faster than they are, such as learning how to use a system.
- They should already know something about your product, such as computers, the Internet, or how their cable TV works.
- They should be comfortable with using electronic equipment, especially since they are purchasing it.
- They should understand your directions the first time without you repeating yourself again and again.
- They are stupid and can't be taught.
- They are too old to learn technical skills.
- They will never change their views.
- They should be excited about the technology and all it can do for them.

It is important to analyze what you expect of your customers and then to let go of these expectations. Remember that your goal is to serve the customers where they are. It may be that you must refer them to a training resource if you cannot reasonably achieve the immediate goals, but do this with patience and respect. No doubt the customer is an expert at something that would baffle you if you were forced to tackle it.

To help develop your patience, try to incorporate the following habits into your day-to-day work:

- Consciously list your assumptions about customers and then let go of them.
- Make it a personal challenge to get each customer to love the products and services they have purchased.
- Smile. It is difficult to be impatient while smiling, and a smile sets others at ease.
- Learn excellent teaching skills so that you can help customers learn how to understand and use their products.
- When your patience is tested, take a deep breath and try to see where your customer is coming from (empathize) before responding.
- Watch your nonverbal responses as much as your verbal responses. Nonverbal cues can demonstrate impatience even more than words.

Critical Thinking Questions

- What "stupid" questions have you heard from customers in your industry?

Try to list at least five. Then, identify the assumption that could be made about the customer and the resulting attitude an employee could have about that customer.

- How might this attitude affect the relationship with the customer?

Open Mindedness

Do you enter into every situation with an open mind? Or, do you try to second-guess the situation? Do you stereotype people based on some criteria (gender, age, dress, speech, etc.)?

Open mindedness means that you are open to new opinions, viewpoints, or ideas of others. Open mindedness is key in the positive attitude required for excellent service. This demonstration of your positive attitude is important for both internal customers and your paying external customers.

A lack of an open mind is reflected in several different common behaviors. Stereotyping, for one, is where one assumes that a person has a specific set of characteristics just because they belong in a specific group. For example, a common stereotype is that older people cannot use or are not interested in technology or that women cannot understand how technology works. This kind of thinking sets up a potentially false situation in your mind from which you speak and act. It might be that the older person is a retired computer programmer or the woman is an aerospace engineer and they both know more about technology than you ever will. When one makes a habit of stereotyping, the subsequent impressions, speech, and behavior are not based on reality and can lead to many negative responses.

Open mindedness is useful beyond interpersonal applications. It is essential as an installer sorts out system problems. Programmers always assume a problem is an installation problem. Installers assume the problem is a programming problem. Both the programmer and the installer will often assume the problem is with the customer. The truth is that the "problem" does not care who thinks what, it exists because something does not work. A core element in service to customers is to fix the problem, not to cast blame. Ultimately, no amount of empathy, patience, or open mindedness will make up for a failure to perform.

Study/Career Tips

We all stereotype people to some degree. Take some time and analyze your own stereotypes. Begin a list of people groups and identify the characteristics that you apply to each group. Becoming aware of your own stereotyping is the first step to getting rid of these assumptions.

Critical Thinking Questions

- What are other stereotypes that you might see in this industry or that you have yourself?
- How might each example potentially affect an employee's dealings with that customer?

Be specific.

Courtesy

What does it mean when someone is courteous? Are you courteous when you deal with others? How specifically do you demonstrate courtesy?

Behaving courteously sends a very positive and powerful message to customers, whether they are your external paying customers or those internal customers with whom you work on a daily basis. Courtesy is also a habit that once formed, becomes second nature.

Characteristics of a courteous employee are reflected in the following behaviors:

- Saying please, thank you, and you're welcome
- Responding with Yes ma'am or No sir
- Saying I'm sorry or excuse me
- Addressing people by their name and using Mr. or Ms. or Miss as appropriate, if you do not know them well

- Saying yes instead of yeah
- Being friendly
- Smiling often
- Opening doors and allowing others to go through first
- Introducing yourself to new people
- Being attentive and focusing on the person in front of you without being distracted
- Using appropriate language.

Courtesy also implies sincerity. Show that you sincerely appreciate your customers by thanking them for their business and being specific. For example: *Thank you for buying our entertainment system. I know you are going to love it. We really appreciate your business.*

Remember your customer's name because you honestly feel that they are important enough to do so. Be sincere in all of your actions. Your sincerity—or lack of sincerity—will be obvious to your customers.

It is easy for people to forget common courtesies when stressed, for example, when a customer is angry or when you are frustrated because of a problem situation. These are the times when you need to be exceptionally courteous. You should be courteous even when you feel that the person does not deserve it and when they are not being courteous themselves. This is why developing the habit of being courteous is so important. If courtesy is a habit, you are less likely to forget about it when stressed.

Study/Career Tips

If you have not developed habits of common courtesy, begin now. Make a list of courteous behaviors and keep that list with you as you go through the day. Look for situations to practice each courteous behavior until it becomes a habit.

Critical Thinking Questions

Make a list of as many different courtesy habits as you can. Strive for at least 25.

- From this list, what courtesy habits do you have right now?
- What habits do you need to develop?

Personal Strategies for Connecting with Customers

There are a number of strategies that you can use to genuinely connect with your customers. Using these strategies thoughtfully and with the goal of meeting the customer's needs and providing them with the best service possible will pay significant dividends in the long run.

Body Language

It is estimated that body language accounts a large part of a conversation. **Body language** is a set of cues that you give to others in the process of communication. That is significant. Understanding another's body language and learning how to adjust your behavior accordingly can go a long way toward connecting positively with your customers. Additionally, effectively using your own body language helps to ensure that you communicate exactly as you intend.

There are many body language cues that can give meaning to your words. In some cases, body language actually changes the meaning of your words—possibly to something very negative. Review these body language indicators.

Important Note: Though body language does make up much of the message, it is not the only part of the message. How you use words is still a critical component. However, when words and body language conflict, typically the body language wins.

Smile

A smile is rarely misinterpreted if it is sincere. A smile reflects a friendly attitude, shows that you are happy to be doing what you are doing, and says that you are eager and willing to help. On the other hand, an inappropriate smile can cause irritation. For example, if a customer is angry and you are smiling, your customer may misinterpret your attitude. You lack congruity, which destroys credibility. Smiling all the time sends the wrong message.

Eye Contact

Making eye contact says that you are interested in the conversation, that you are giving the person your full attention, and that you are confident and honest. Looking away or downward tends to reflect insecurity, boredom, or dishonesty. When outside in the sun, many people wear sunglasses. This typically makes many people uncomfortable. Take off your sunglasses, even if it is for a short time, to reassure your customer that you are interested and honest.

Good Posture

Good posture includes standing up straight with shoulders back and head up. It is at the same time relaxed and natural. Good posture reflects confidence and competence. Maintaining good posture is a habit. How is your posture? If it needs work, try to focus on it throughout the day so that you can develop the habit. Good posture is not only good for how you impress customers, it is also good for your health.

Hand-to-Face Movements

Holding the chin, scratching the face, or tugging at the ear is often a sign of indecision or further questions. If you read this in your customer, you might try to anticipate the questions and answer them as best you can. If the customer reads this in you, they may lose confidence in your ability to solve the problem or provide the service. Touching the nose often indicates untruthfulness. Touching or rubbing the eyes often indicates disbelief or doubt.

Arm Movement

Folding the arms across the chest can indicate that a person is defensive. For example, if you present folded arms to your customer, they may perceive you as hiding something or resistant to what they are saying. If your customer begins to unfold their arms, they may be warming up to what you are saying.

Displacement Activities

Displacement activities are those that relieve stress or anxiety: clicking a pen, biting nails, tapping a table, bouncing a knee, chewing gum, playing with hair, kicking the foot, etc. Be aware of these actions and eliminate them. These behaviors signal nervousness and insecurity and are very annoying.

Important note: Body language cues sometimes mean different things in different cultures.

Tone of Voice

Sometimes how you say something means as much as what you say. The wrong tone can cause a misinterpretation of your words. Saying *thank you* in an angry tone only serves to agitate your customer. Asking about the problem using a disinterested tone shows the customer that you are not sincere. Using sarcasm and a sarcastic tone typically causes a customer to become angry and to feel disrespected.

Combining a positive, friendly, and confident tone with positive and confident words such as absolutely, definitely, not a problem, and so forth can be very effective. Say the following phrases with a positive, friendly, and confident tone to get the point:

Absolutely. I can have this fixed in no time!

Definitely. I will order the part for you today.

Not a problem. I will reschedule the service call to that date.

Yes! I will be happy to move this for you.

Try to match your tone to the customer's needs. If the customer is in a hurry, then make your tone urgent and energetic. If the customer is frustrated, use a confident and helpful tone. If your customer is doubtful or has many questions, use a reassuring and confident tone.

Attentiveness

Paying attention to customers is critical. Being attentive is demonstrated by many different behaviors. First, greeting customers by their name in a friendly, eager, and energetic manner starts off the interaction on a positive note. After the greeting, being attentive means to give your undivided attention to the customer. Examples of divided attention include talking with coworkers about issues unrelated to the customer, answering a cell phone, looking out the window or through notes or files, fiddling with equipment, and so forth. To be attentive to the customer means to first listen to what they have to say and *then* go about completing your tasks. Attentiveness also involves prompt response to calls or e-mails. It means prompt attention to issues, questions, or tasks.

Being attentive to the customer also means paying attention to details—the details associated with every step of the process from the products themselves to how they are delivered, installed, maintained, used, serviced, billed and paid for, connected to other services or products, and so forth. Think through your products and services and list as many different details about each step as you can. There should be hundreds of items on the list. Then, think through each of these details to see if it is good for the customer or if it poses a confusion or inconvenience for the customer. Finally, fix the issues when you can. This exercise will help you increase the attentiveness to your customers.

Finally, attentiveness includes behaviors that might fall into the category of thoughtfulness. For example, a system installer who must come into a customer's home is attentive to the customer needs and to details if he puts protective covers on his boots or on the carpet and if he thoroughly cleans up any mess he might have made during the job. Another example is to ensure that the customer knows how to use every aspect of a piece of equipment before leaving. One sales specialist in this industry was closing a huge project when the customer explained

FIGURE 2–2 Being an effective listener will foster an effective and productive relationship between you and the customer.

why she would not consider one of the competitors. In another home, the competitor's installers left small bits of wire and insulation on the floor and carpet. Her children were young and one of the babies was putting those pieces in his mouth. The competitor lost a project worth several hundred thousand dollars because of this simple lack of attention to details and the customer's needs.

Listening

There is a significant difference between listening and hearing. To listen means that you truly attempt to understand what the speaker is saying. Real listening is a highly active process. Without listening, there is no communication—only speaking and hearing.

Effective listening does several things for the relationship:

- It shows that you sincerely care about the customer and the customer's needs.
- It demonstrates attentiveness to the customer.
- It allows you to gain critical information with which to complete your task successfully or to solve the problem.
- For frustrated customers, it reduces irritation by ensuring the customers that they are being heard.
- It fosters an effective and productive relationship between you and the customer.

Chapter 3 discusses some important strategies for effective listening. Take some time to critically think about your listening skills and work diligently to perfect them so that you can develop long lasting and effective relationships with your customers.

Responding

How you respond to the customer typically determines how well your service is received, which ultimately translates to either customer satisfaction and loyalty or to customer dissatisfaction.

There are several ways to respond effectively to customers that will help to make them feel served and fully attended to.

Acknowledge the Customer's Needs

If the customer sees that you are trying to understand what they want or need, they typically will be satisfied, even if you cannot deliver. It is important to show that you understand what they are saying and what they want.

I understand that you would like this connection moved to the other side of the room.

Ask Questions Rather Than Making Statements

No one likes to be ordered around. To avoid the appearance of barking orders, ask a question instead.

Would you like to set up a meeting with the contractor to get his advice on these issues?

Rather than

Call the contractor. We need his advice.

Focus on the Problem Rather Than the Customer's Emotions

When dealing with an angry or highly emotional customer, maintain your focus on the problem rather than on the customer's behavior. Typically, when the behavior becomes the focus, the real problem does not get resolved.

Ask Questions to Get More Information

Keep questions simple and ask only one question at a time to avoid confusion. If you must ask a series of questions or if the question is complicated, break it down into smaller bites. Also, try phrasing your questions differently. Use "*Tell me about . . .*" or "*Describe . . .*" or "*Give me more information on . . .*".

Admit Any Mistake You Make and Apologize Sincerely

Typically, people are forgiving of mistakes if the offender confesses, apologizes sincerely and immediately, and tells the customer how it will be remedied. It is difficult to stay angry when someone owns up to a mistake and fixes it. After you admit the mistake and apologize, refocus attention to how the problem will be fixed. Never make an insincere apology. The customer will see through insincerity easily.

Explain the Actions You Are Going to Take

Information goes a long way toward satisfying a customer. If you are going to complete a job in a customer's home, explain what you will be doing and why in terms that the customer can understand. If you are checking on the solution to a problem, explain to the customer how you are going about finding the answer.

Find Things on Which to Agree

Agreeing with the customer on even small points often creates the perception that you and the customer are on the same side. This works especially well for angry or dissatisfied customers. For example, if a speaker does not work as it is supposed to, simply saying, *"You are right. This just isn't how it is supposed to sound,"* may immediately reduce the customer's anger and create a positive situation in which you can take the steps necessary to fix the problem. There may be other elements of the problem on which you can easily find agreement. Use these agreement points to gain your customer's trust and cooperation.

Avoid the Words "I Can't."

Focus on what you can do rather than what you can't do. If you cannot do exactly what the customer wants, explain what you can do for the customer that either comes close to the customer's request or meets the same need in a different way. Be a problem solver.

To connect with customers, develop your skills in and practice the following strategies:

- Evaluate your body language and use it to convey the appropriate messages.
- Focus on how you say things and your tone of voice.
- Be attentive to customers.
- Develop effective listening skills and practice these with customers.
- Respond positively to customers.

Understanding Customer Needs and Expectations

Using theses strategies, you should be able to connect with your customers and understand their needs to determine exactly what the customer expects of you, your business, and your products and services.

As a reminder, here are some fundamental strategies that you can use together to ensure you understand your customer's needs and expectations:

- Listen intently and actively to your customer.
- Summarize what you think your customer is saying.
- Ask questions to clear up anything you don't understand.
- Use appropriate body language and nonverbal cues to ensure that you are sending the message that you value your customer.

Be attentive to your customer both in immediate interactions and any tasks that you are to complete.

Service to Internal Customers

As explained in Chapter 1, internal customers are those individuals and departments who rely on you and on whom you rely to ultimately meet the needs of your external paying customers. They are your coworkers, members of other departments (sales, advertising, production, accounting, human relations, management, etc.), and other service providers (contractors, builders, architects, electricians, carpenters, designers, plumbers, security system technicians, cable providers, telephone technicians, and so forth). Without these individuals and departments, your paying external customers will not receive the products and services they pay for.

Think about your current or future position in this industry.

- If you quit working for two weeks with no replacement, who would complain?
- What specifically would they complain about?

Make a list of as many people, departments, and partners that you can think of. Then identify exactly what they would complain about if you did not do your job.

Critical Thinking Questions

Successful service to internal customers requires effective communication, excellent teamwork, and an appreciation and value of everyone involved in delivering and servicing the products to the customer. Let's look at some of the important elements required for excellent service to internal customers.

Excellent Teamwork

A **project team** is any group of individuals who interact to achieve a specific purpose, such as installing a home entertainment system in a newly constructed home. In any industry, there are many types of both informal and formal teams. You participate in a project team, whether it is structured or not. It takes a team to successfully install a system in a customer's home. It takes a team to sell a product and provide the

FIGURE 2–3 The total success of a project depends on the effective collaboration between all members of the project team.

appropriate services to a customer. A project team might include the owner, architect, electrical contractor, builder, installation technician, and sales person responsible for the product. You may not physically work with each member of your team; however, the total success of the project depends on the effective collaboration between all of the team members.

There are many advantages to looking at each project as a team project.

- Better and faster communication
- Improved problem solving
- Better decision making
- Improved performance
- Improved morale
- Higher customer satisfaction.

There are many important behaviors to being an excellent team player.

- Be accountable
- Be reliable
- Communicate constructively
- Listen effectively
- Actively participate
- Share with team members
- Cooperate with team members
- Be flexible
- Help to solve problems
- Be supportive of team members
- Be committed
- Avoid gossip
- Compliment your team members at every opportunity
- Take action
- Be proactive
- Give credit to others when appropriate
- Say thank you
- Show appreciation
- Resolve conflicts immediately
- Use positive language
- Avoid negativity.

Critical Thinking Question

- How does poor team work affect the ability to provide excellent service?

List as many different examples as you can.

The Teamwork Perception

Your external customers will see whether or not you have excellent teamwork in your company. If they perceive that you and your coworkers and partners work well together, they will have more confidence in your company's ability to provide them with quality products and services. If they see that there is little or no teamwork or that you and your coworkers and partners do not work well together, they will lose confidence in the company and its products and services.

There are many behaviors and strategies you can incorporate into your daily work that foster excellent teamwork and demonstrate to customers that you value those around you.

- Never complain about anything work related within earshot of a customer (internal or external). It says that there is something wrong internally with your company.
- Never complain about another company, coworker, or department to an internal or external customer. It says that your company doesn't get along well with others.
- Never talk about external customers to other customers (internal or external). It makes people wonder what you are saying about them to others.
- Whenever possible, compliment others in your company or a partner company in front of customers (internal or external). It makes them look good and shows that you value those with whom you work. Consider telling their supervisors as well.
- Say thank you. It shows that you value those with whom you work and makes a very positive impression on external customers. Thank you's are also contagious.
- Whenever possible, show genuine appreciation to those with whom you work. It makes them feel valued and will come back around to you eventually.
- Whenever possible, help out a coworker, partner, or department, even when it is not your job. It shows that you value the other's work and will help you to better understand what they do. This too will come back around to you eventually.
- Discuss processes and projects together to learn from both successes and mistakes. Strive never to make the same mistake twice and to create streamlined processes that work for everyone involved.

Summary

In this chapter, we discussed some important characteristics of an excellent service attitude: confidence, competence, appreciation, empathy, honesty, reliability, responsiveness, patience, open mindedness, and courtesy. This chapter addressed personal strategies employees can use to connect well with customers such as body language cues, tone of voice, attentiveness, effective listening, and responding well to customers. The chapter also discussed the importance of excellent teamwork and the characteristics required for effective project team members.

Important points in this chapter include:

- Attitude is everything when it comes to dealing with customers.
- There are several important characteristics to making up an excellent service attitude: confidence, competence, appreciation, empathy, honesty, reliability, responsiveness, patience, open mindedness, and courtesy.
- A high level of confidence lowers stress and anxiety both for yourself and for those around you.
- A high level of competence increases confidence and should be an ongoing goal for employees.
- Life-long learning is essential for career success over the long term.
- Appreciation of yourself and your customers is a mindset and is established through self-talk.
- Showing empathy to customers is an essential behavior for excellent service, especially when customers are upset, angry, or frustrated.
- Honesty is an essential element in business regardless of the level of employee. There are several ways to avoid "honesty" problems with customers.
- It is essential to be reliable when working with both internal and external customers.
- Being responsive to customers is a significant characteristic of employees who provide excellent service.
- Employees should be patient with customers even when it is difficult to do so.
- Maintaining an open mind helps to ensure that you do not make inaccurate assumptions about your customers.
- Courtesy is essential in excellent service and should be practiced at every opportunity.
- There are many body language cues that make an impression on customers and to ensure effective communication. Body cues include smiling, eye contact, good posture, hand-to-face movements, arm movements, and displacement activities.
- Appropriate tone of voice, attentiveness to customers, effective listening, and responding appropriately to customers are keys to excellent service.
- Workplace projects should be considered to be team projects and all employees should develop essential project team skills.

Key Terms

Body language Body language is a set of cues that you give to others in the process of communication.

Empathy Empathy is putting yourself in the other person's shoes.

Life-long learning Life-long learning is the idea that a professional must continue developing knowledge and skills throughout their life in order to maintain competency for the workplace.

Project team A project team is any group of individuals who interact to achieve a specific purpose.

Self-talk Self-talk is what you say silently to yourself as you go through the day or are faced with difficult situations.

Review Questions

1. Why is attitude so important in providing excellent service? List as many specific reasons as you can. Be sure to address both internal and external customers.
2. How specifically does a high level of employee confidence impact the employee's internal and external customers?
3. How might an employee's self-talk affect the ability to provide excellent service?
4. Why is it important for an employee to have empathy with both internal and external customers?
5. How can employees of a company demonstrate honesty to customers? Be specific. What are

potential situations in which "honesty" problems commonly occur?

6. Why is reliability so important in providing excellent service to both internal and external customers? List some specific examples.
7. What does it mean to be responsive to a customer? List some specific examples of an employee being responsive.
8. What specifically should an employee do when a customer is testing his patience?
9. How does the lack of an open mind impact an employee's ability to provide excellent service?
10. What are ten examples of courteous behavior?
11. What are six specific body language cues that tend to make a positive impression on a customer? Explain the reasoning behind each cue.
12. What are six specific body language cues that tend to make a negative impression on a customer? Explain the reasoning behind each cue.
13. How can an employee's tone of voice affect an interaction with a customer?
14. How does an employee's attentiveness to a customer impact the quality of service? Why?
15. Why is it so important to develop excellent listening skills in order to provide excellent service? Be specific.
16. What are eight strategies for responding to a customer?
17. What are specific ways that employees can ensure they understand their customers' needs and expectations?
18. What behaviors are associated with being an effective project team player?
19. What are advantages of looking at every project as a team project in this industry?

Effective Communication for Excellent Service to Customers

OBJECTIVES

After studying this chapter, you should be able to:

- Define the characteristics of clear communication in face-to-face interactions, on the telephone, in e-mail, and in written documents.
- Define the characteristics of an effective listener.
- Explain ways to ask questions for better understanding.

OUTLINE

Introduction

Think carefully about how you like to feel when interacting with an employee of a company. What specifically makes you feel this way? Is it the personality of the employee? Is it what they say or do? Is it how they treat you? Since most of our experience with employees of a company is in some form of communication, it is essential to understand how to make every communication an opportunity for excellent service. This chapter discusses some simple strategies that can be used for face-to-face communication, communication on the telephone, communication through e-mail, and communication through the various written documents commonly used in the industry.

Clear Communication

Clear communication has a beginning, a middle, and an ending, regardless of the situation in which the communication occurs. By thoughtfully considering how you handle the beginning, middle, and ending, you are more likely to clearly communicate and leave your customer feeling valued and well-served. Let's look at effective strategies for the beginning, middle, and ending in several different types of communication situations or communication mediums.

The Communication Process

Effective communication is complicated and sometimes difficult to achieve. A careful study of effective communication reveals that there are four main steps commonly called the **communication process.** Regardless of the type of communication medium you use, the process is the same. There are numerous ways for the communication to be derailed resulting in no communication at all or a miscommunication. Let's review each step in the communication process (Figure 3–1) in more detail.

Critical Thinking Questions

- Whose responsibility is it for effective communication?
- What kinds of things can go wrong in communication?

Make a list of as many barriers as you can.

Encoding and Sending the Message

Figure 3–1 explains a simple communication process. A sender must first encode the message before sending it. Encoding means that the sender must verbalize the message or write the message down for a written communication or an e-mail message. To encode an effective message, you must think very carefully about the purpose of the message and the result that you want. You must encode the message in a way

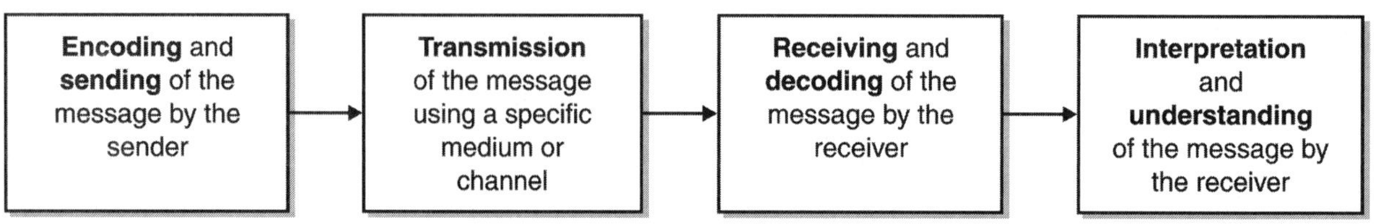

FIGURE 3–1 The Communication Process

that you know the receiver will hear and understand. For example, you would not encode the message using Spanish words if you know the receiver does not understand Spanish. You would also not use highly technical jargon if you think that the customer might not be familiar with the technology. Additionally, you must think of the impression that you want to make on the receiver with the message. Part of the message might include facial expressions, nonverbal behaviors, gestures, professional appearance of a written document, tone of voice, visual aids, and so forth. Remember that the message is much more than words or text.

Transmission through a Medium or Channel

Next, the sender must send the message using an appropriate **communication channel** or communication medium. A communication channel reflects the way you send the message: written format, telephone, face-to-face communication, etc. You should select the best channel for the purpose of the message as well as the most appropriate channel for the receiver. For example, you would not send an e-mail to a customer who does not often use e-mail, and you might be required to use the telephone because your customer is not available face-to-face. For some problems, it might be best to use face-to-face communication. A written letter might be the best when written documentation is required for future reference. A voicemail message might be appropriate for simple conveyance of information, such as confirming an appointment or providing ordering information. Think about the goal of the communication and choose the channel thoughtfully.

Related to channel of communication is the venue of the communication. An effective communicator must select the venue of the communication carefully for each situation. Depending on the stakes and the degree of trust and familiarity already established with your customer, your venue for communication could range from a simple e-mail to a dinner meeting. It is up to you to determine what is at stake and what is the likelihood of misinterpretation of the message. As a rule, the higher the stakes or the weaker the familiarity with the recipient, the riskier an e-mail becomes. We have all experienced an unexpected reaction to what we thought was a perfectly innocent e-mail. One suggestion would be to combine two technologies: you could precede an e-mail with a voicemail so the recipient will (1) expect the imminent e-mail and (2) experience the tonality of your voice and determine the mood of the message. For sensitive or problematic situations, a face-to-face interaction is better than written channels.

Receiving and Decoding of the Message by the Receiver

Finally, the message must be received and decoded by the receiver. This means that the receiver must be able to hear or read the message clearly. Barriers to receiving the message can be external such as environmental noise, static on a cell phone, poor Internet connection, cutoff faxes, lost notes, and so forth. Internal barriers include distractions, stress, insecurity, and other influences. There may also be barriers caused by intense emotion such as frustration, anger, or fear. It is the responsibility of the sender of the message to ensure that the receiver actually receives the message. There may also be barriers in the decoding of the message. For example, there may be a language barrier. A receiver who speaks a different language or cannot hear well may not understand the message. A receiver who is not familiar with technology may not understand the technical terms and concepts.

Critical Thinking Questions

Think about your job or the job you soon hope to have.

- What barriers to receiving information might your customers experience?

Try to think of specific scenarios and be as specific as possible.

- What can you do to overcome these barriers?

Interpretation and Understanding of the Message

When receivers decode the message, they will interpret it using all of the information (verbal and nonverbal) sent by the sender. They will also apply their own experiences, background, and attitudes to that message. A customer's biases or assumptions also impact how the message is received. For example, a customer who has a bias against women as technicians in a high technology area may not be open to a female technician's advice or information. There are many other types of biases. Though you probably cannot change that bias and it is definitely not appropriate to try, it is important to be aware that bias and other incorrect assumptions do occur.

Critical Thinking Questions

- What are effective, tactful, and professional ways to handle a customer who has an obvious bias toward you as an employee?
- What are ineffective and unprofessional ways to deal with this kind of situation?

Misunderstanding occurs when the sender has not accurately sent the message, the message was sent via an inappropriate channel, or the impression left by the message was misinterpreted. As you can see, the majority of the responsibility for communicating an effective message lies with the sender.

Communication Channels

There are four important communication channels that are commonly used with customers: face-to-face communication, the telephone, e-mail communication, and written documents. For each channel, there are several strategies that can help ensure that you provide excellent service. To make things simple and organized, let's break down each type of communication into a beginning, a middle, and an ending.

Face-to-Face Communication

In face-to-face situations, your customer both sees and hears you. The impression you make on your customer is influenced by how you look, how you speak, your body language, your words, and how you behave. It is important to be aware of all of these aspects of your message and channel to effectively communicate. On the other hand, the advantage of face-to-face communication is that you can see the customer, which allows you to use the customer's body language and behavior in addition to the words they use to guide the communication.

The Beginning

The beginning of a face-to-face communication starts even before any words have been said. Remember that people see you first and hear you second. It starts with the customer's very first impression of you. Your appearance, demeanor, and approach set the stage for how well you communicate with the customer from then on. The saying "first impressions are hard to change" is very true in providing excellent service. Once the customer has developed an impression, they view everything you say and do with this impression in mind. Your customer also judges your entire company based on these first impressions. Remember too that your customer represents a potential endless stream of referrals for your business. It is critical to get first impressions right.

There are many strategies you can use to ensure that the first impressions you make on the customer work to your advantage rather than hinder future communication.

Maintain a Profession Appearance. A professional appearance does not mean wearing a suit and tie (unless the situation calls for a suit and tie); however, it does mean dressing in an appropriate manner for the job you are to do. Regardless of the task you are to perform, maintaining a clean, well-groomed, healthy, and put-together appearance goes a long way to influencing your customer's first impressions. Good grooming means clean hands and finger nails; clean and styled hair; clean and groomed face, body, and teeth; clean and well-fitting clothing that is not in any way offensive to a customer; and so forth.

Maintain a Positive and Inviting Demeanor. Even if you look good, a negative demeanor can ruin a first impression. Your demeanor is the way in which you behave. What does your demeanor say about you? Are you interested, enthusiastic, happy, knowledgeable, glad to be there, confident, and friendly? Or, are you bored, angry, unhappy, tired of your job, insecure, and unpleasant to be around? Your demeanor is determined by your body language, smile, eye contact, energy level, and enthusiasm.

Greet the Customer Enthusiastically. After your appearance and demeanor, first impressions are formed by the first words you say and how you greet your customer. If you are meeting a customer or contractor for the first time, introduce yourself and identify your company and your place in it.

"Hi, I am Michelle Morrison. I am a technician for ABC Company and am here to install your system."

Make an opening statement that says that you are glad they chose to do business with your company.

"I am excited you chose this system. I think you will really enjoy it."

If you know your customer's name, use it.

"Hi Jim, great to see you."

Make good eye contact, smile, and demonstrate your genuine enthusiasm for your job.

Pay Close Attention to Your Customer. Try to determine the demeanor of your customer. Is she upset or angry or disappointed? Is he apprehensive about his purchase? Is she excited about her new system? If possible, try to anticipate your customer's concerns or problems and alleviate them.

Examples are:

I know it is raining outside. I promise I won't drag any mud into your beautiful home; I will be very neat.

This may look like a big job, but I work quickly. I'll be done in no time.

The Middle

We have discussed many strategies for how to treat a customer in order to create customer loyalty and satisfaction. Remember to use all of these strategies to demonstrate your positive attitude, to ensure that the customer feels valued and appreciated, and to help solve their problem or meet their need.

There are some other important strategies you can use to ensure that the whole encounter is positive.

Be Attentive. Be sure to focus on your customers and their needs. Avoid being distracted by anything else going on around you or by your own internal thoughts. Review the keys to attentiveness in Chapter 2 and make these strategies habits in your everyday dealings with customers.

Show Your Customers. A large number of people in the world are visual learners. This means that they process information visually and like to see things as they hear the information. Use this to your advantage. If you are explaining an installation, take the customer to the site of the work and show them what you are planning to do. If you are solving a technical problem with a piece of equipment, open up the equipment and show the customer the defect part or give them a mini-lesson on how the equipment works and what went wrong. As with all communication with customers, use language and terms that they can understand. If your customer must choose from several options, bring in the options, if possible, or show pictures of the options while explaining the differences. If your customer must decide where to place some part of the equipment, walk around the site with them as you explain the advantages of one site over another. If actual products or pictures are not possible, consider drawing a simple diagram of your plan or the product as you explain it using very descriptive language.

The Ending

How you leave a customer helps to solidify your relationship and to assure the customer that you have successfully completed your task or solved the problem. It also provides an opportunity to thank the customer for their business and show your appreciation.

There are several strategies that you can use to successfully end a face-to-face communication with a customer.

Keep the Customer Updated. Most projects last for weeks, some last for months or even years. Whenever you leave the project tell the customer or contractor (1) that you are leaving, (2) what you accomplished, (3) what you did not accomplish, and (4) when you are returning.

Finish the Job Completely. For tasks that you perform, be sure to clean up any mess you may have made, put things back where you found them, and make the products look like new. Take out the trash you brought in rather than filling up the customer's trashcan. Clean off smudges from a wall or facing or organize equipment cables behind a desk. Go the extra mile. You may need to ask the customer how they want things to look or if you can help move furniture back to its original place. Offer to vacuum the carpet where you worked if needed. Even small gestures go a long way to customer satisfaction.

Ensure that the Product or Service Works. There is nothing more frustrating than for a customer to try to use their newly purchased product to find that it does not work. Before a job is complete, ensure that everything functions as it should. Test the entire system yourself before declaring your job done. Also test existing systems you may have been peripherally involved in. For example, you could leave a customer without phone service because you connected something and left it there. This is similar to a surgeon who leaves a sponge inside a patient. Also avoid the practice of fixing the first thing you find wrong and then assuming you are done. Check the system entirely.

Make Sure the Customer Knows How to Use the Product or Service. When necessary, walk the customer through the use of the product starting with how to turn it on. Make sure that they understand the basic functionality. Show them how to shut it down, if applicable.

Chapter 5 provides many strategies for educating the customer on products and services.

Make Sure the Customer Is Satisfied. You have not completed your job until the customer is satisfied with it. When you leave a customer, provide them with a summary of what they purchased, the service you provided, and how you solved the problem, if applicable. Then, ask them directly if they are happy with what has been done or what they purchased.

Thank the Customers for Their Business. Always let your customers know that you appreciate their business. Remember that you always speak on the behalf of your company. If you are in a situation where the customer did not give you their business, thank them for considering your company and offer to help them with something else in the future.

Look for Opportunities to Make Additional Sales. Though your job may not be sales related, you do represent the company and may be the only employee to have direct contact with the customer. Look for opportunities to sell additional products or services to your customer. Obviously, this strategy must be used tactfully; however, both you and your company will benefit as more products and services are sold. In order to know when this kind of opportunity exists, you must know the products and services of your company, even if they are not directly related to your job. You should understand what your company does, keep up with new services or technologies that it offers, and have a working knowledge of the products and services.

Study/Career Tips

Customers sometimes write notes to an employee's supervisor to tell them of either excellent service or disappointing service. Assume that one of your customers has written a note to your supervisor to tell him what they think of their first experience with you. What would you want it to say? Write down what you would want the note to say in detail. Be specific. Now, are there things that you need to work on to make that note reflect reality?

The Telephone

Every time you make or receive a telephone call, you represent not only yourself, but your company as well. It is estimated that over half of the communication in a face-to-face interaction occurs through body language. Effectively dealing with customers on the telephone is more difficult than in a face-to-face interaction. You do not have the advantage of using body language, facial expressions, and eye contact to judge your customer's response and understanding. You must rely only on the words and tone of voice. The customer cannot read you as well either. The telephone presents many barriers to effective communication;

however, telephone communication may establish the first impressions of you and your company for the customer. It is important to master excellent telephone etiquette skills in order to provide excellent service.

As with face-to-face interactions, a telephone conversation has a beginning, a middle, and an ending. There are several strategies, however, that you can use to overcome the barriers and ensure that the communication is effective.

FIGURE 3–2 Effective communication, whether on the telephone or face-to-face, is important in building positive customer relationships.

The Beginning

Before even answering the telephone, consider that the caller may be one of your customers. Answer your phone, if at all possible, rather than letting a machine answer it. From the second you pick up the receiver, you begin to make an impression on the caller.

Focus on the Call. Answer the phone promptly (on the first or second ring, if possible), identify yourself to the caller, and give your customer your full attention. Stop any other conversation before you answer the phone. Do not chew gum, drink, or eat while talking on the phone.

Identify Yourself and Company. Answer the call by identifying your company and yourself. If your company has a specific way to answer the phone, follow this policy. If not, consider answering the phone with your company names followed by your name and a greeting. Remember that your very first statement makes a first impression that will be carried through the call.

Good morning, ABC Entertainment Systems. This is Tom speaking. How can I help you today?

If your customers call you on your personal phone, consider answering your phone at least with your name.

Good morning, this is Tom.

Be Enthusiastic and Welcoming. Remember that the call may be the only chance you have to make a positive impression on your customer. If the caller introduces herself, repeat the name (e.g., *"Hi, Jamie. How are you?"*) and use it often in the conversation.

Be Clear in Your Speech. Speak slowly and clearly. Spell out names that might be difficult to pronounce, understand, or spell.

Pay Attention to the Quality of Your Voice. Quality involves rate, pitch, volume, clarity, and tone. Fast talkers are often perceived as too busy to talk or untrustworthy. Slow talkers may be perceived as mentally slow. Both high-pitched and low-pitched talkers are often irritating. Loud people are often considered to be brash or overbearing, while very soft talkers are perceived to be meek or insecure. Clarity means that you articulate your words. This presents a more professional impression. Finally, your tone should sound enthusiastic and expressive.

Transfer Calls Appropriately. When transferring a call, explain to the caller where you are transferring them and why. If the caller has reached you by mistake, try to find out what they need and ensure that they get to the right department.

Critical Thinking Question

- What are some ways to answer a phone call that would turn off a customer immediately?

List a few. Think about both an employee answering a company phone and that person answering a personal cell phone.

The Middle

As with face-to-face interactions, your attitude is reflected in how you treat your customers. And, although many of the nonverbal cues are not available in a telephone call, a customer-focused attitude can still shine through.

Maintain an Excellent Service Attitude. As with a face-to-face interaction, try to determine the question, request, or problem. If the caller is upset, empathize with them without becoming defensive. Try to understand their issue thoroughly and acknowledge it directly. If needed, apologize for any problem on behalf of your company. An apology goes a long way toward establishing a positive relationship. Most customers are very forgiving if they know the problem will be remedied and that the company has taken responsibility for it.

Place the Caller On Hold If Necessary. Always place a caller on hold so that they do not accidentally hear a conversation going on in your office. If you must leave the line for a long period, give the caller the option of waiting on hold or receiving a return call. When you do return to the caller, thank them for waiting. Give the caller an update on your status every 30 seconds or so to assure the caller you have not forgotten them.

Be as Helpful as Possible. Assure the customer that you can help them. Reply to the customer's opening statement with *"You bet. I can help you do that."* Be as helpful as you can and handle as much of the problem or request as you can. For example, complete as much paperwork as you can, find the correct sizes or order numbers, or take down as much information as is feasible to save time later on and to assure the customer that they are being served.

Avoid Negative Phrases. There are many negative phrases that you should avoid or rephrase to more positive tone.

> *We can't do that.* Rephrase to *That might be tough. Let me see what I can do about this.*
>
> *You will have to . . .* Do not sound accusatory. Rephrase to *Here is our procedure on . . .*

No. Never start off a sentence with a negative. Rephrase the sentence to start with what you can do to solve the problem.

I don't know.
This statement sounds like you do not know what you are doing. Tell the customer what you do know and that you will find out any other information you need.

Tell the Customer What You Are Doing Throughout the Call. Be sure to let your customer know what you are doing throughout the call. In some cases, you may have to stop talking to take notes or view information on the computer. Tell the customer that you are writing or typing in the information or waiting for the computer to respond. Do not leave the customer in silence, wondering if you are still there. Talk to the customer while waiting on your computer. This downtime is an opportunity to continue to develop a good rapport. If you need to ask a series of questions to get the needed information, let the customer know this so that they do not feel as though they are being interrogated.

Study/Career Tip

What might customers think if you just remain silent while working on their problem on the phone or while waiting for a computer to bring up information?

Do Not Take Things Personally. When customers are unhappy, they can overreact, behave inappropriately, or be rude. Do not overreact yourself or take what they say personally. Chances are you did not cause the problem; however, you still must represent your company to the customer during the call. Respond calmly, professionally, and in a helpful manner and be sure that your words and voice tone relay this demeanor. Do what you can to help the customer. More will be said about handling difficult customers in Chapter 5.

Take Good Messages. If you are taking a message, be sure to write clearly and take down all the important information: name, telephone number, time and date, purpose for the call. Be as detailed as necessary. If you are answering the phone for someone else, do not tell the caller personal information such as why the other person is unavailable or details that may make the person look unprofessional. Instead, give general information that gives a professional impression. Figure 3–3 suggests appropriate responses for common situations.

Transfer the Call If Needed. Learn how to use the various functions of your telephone system, especially the sometimes-complicated process of transferring a call. Practice this process so that you do not disconnect your customers. When you do transfer a call, tell the customer that you are going to transfer them and tell the receiver of the call who is calling and what they want. This means that you need to know the customer's name and the purpose of the call. If you can, stay

When you mean...	*Say...*
John is sick today.	John is not available. May I take a message?
I don't know where Martha is.	Martha stepped out of the office. Would you like her voicemail?
Ed isn't in yet.	I expect Ed shortly. Could I take a message?
Trisha left early today.	Trisha is not available right now. Would you like to leave a voicemail?
Mike is too busy to talk to you	Mike is not available right now. May I take a message?
May is in the restroom.	May stepped out of the office. May I take a message?

FIGURE 3–3 Message Responses

on the line until the person answers to be sure the customer does not get passed around or disconnected.

The Ending

The last few seconds of the phone call allow you to leave the customer with a positive impression that will last until the next communication. It is important to bring a positive closure to your telephone communication.

Confirm What You Are Going to Do and/or Confirm Important Details for the Customer. Confirm any details such as times and dates or the details of an order. Ask the customer if the information is accurate or if the solution is satisfactory. Or, ask if you have handled the situation satisfactorily. Also ask if there is anything else that you can help with while on the call.

Be Sure the Customer Has Your Name and Contact Number. Give your name again to the customer along with a way for them to contact you. This gives the message that you are taking responsibility for the agreed upon tasks or information. Do not depend on the customer remembering your name from the beginning of the call.

End on a Positive Note. Thank your customers for calling or taking the time to speak with you. Thank them for their business. Tell the caller you have enjoyed speaking with them.

Say "Goodbye" in a Friendly Way. Say goodbye to the caller and thank them for the call or for their business. Allow the caller to end the call.

Follow Through on Promises. As with any other communication with customers, it is essential to follow through on promises made during the telephone conversation. If you promised to find out more information and then call the customer back, do so as quickly as possible. If you promised to take an order or message, do so accurately.

Outgoing Calls

There will no doubt be times when you must call the customer yourself. When doing so, consider the following tips.

Greeting

Greet the customer in a friendly and enthusiastic tone and identify yourself and your company immediately. Be sure that you are speaking with the appropriate person. State why you are calling and ask if this is a good time to complete your business.

Prepare

Prepare what you are going to say before making the call. Make sure you have all the information you need so you do not waste your customer's time. Make notes for the call.

Practice

If needed, practice the message. This will help if you have to leave a message on a machine.

The Voicemail Message

There will no doubt be many instances in which you cannot answer the phone and your voicemail will make the first impressions for you. Consider these tips for making a great impression on your voicemail.

Record a Personal Greeting

Record a personal greeting rather than using the standard greeting on your voicemail service. In doing so, you provide a personal touch rather than leaving the customer feeling frustrated. Also, create the message yourself to give the caller a sense that you listen to your messages yourself.

Include Personal Information

Include your name and title or department to assure the caller that they have reached the right person. Also, include in your message your normal working hours or times that you can be reached. If someone else can help the caller in your absence, provide the information on how to reach this person. Similarly, if you have a cell phone or alternative number, include that on the message.

Change Your Message Appropriately

Change your message accordingly if you are on vacation or have different hours temporarily. This will tell the caller not to expect a call back until you return and prevent them from being frustrated.

Check Messages Often

Check your messages often and return calls within 24 hours. Sooner is better. Respond to your messages as quickly as possible.

Critical Thinking Question

What are the potential consequences with an inappropriate message on an answering machine or one that provides little information? List as many ideas as you can about what might be your customer's response.

Leaving a Message

There may be instances when you call a customer and must leave a message. Use these strategies to make a positive impression with your customer, even on their answering machine.

- Be friendly and enthusiastic.
- Introduce yourself using your name, your title, and your company.
- Speak slowly and clearly.
- Leave the time and date that you called.
- Leave a callback number.
- Leave the time that you can be reached.
- Consider repeating the phone number to assure that the caller heard it correctly.
- Explain the purpose of your call and leave any important details.
- Specify what you want the customer to do.
- Do not leave personal information on a message machine.

E-Mail Communication

Today, many communications with coworkers, contractors, and other customers are made through e-mail. Since e-mail communication has neither the advantages of verbal and nonverbal cues found in face-to-face interactions nor the advantages of voice quality and immediate two-way communication over the phone, effective communication via e-mail can be tricky.

Netiquette is a relatively new term meaning network etiquette and other appropriate technology use. Basically, netiquette means the manners or professional courtesies followed in an on-line communication. For excellent service, it is essential to hone your skills of communicating online. As with face-to-face communication and telephone conversations, e-mail communications also have a beginning, a middle, and an ending. Consider composing your message offline to avoid accidentally sending an incomplete message or one that you have not reviewed.

The Beginning

With all forms of communication, the way you present yourself to your customer establishes an impression for both you and your customer.

In order to make a positive and professional impression, there are several rules that should be followed when using e-mail.

Do Not Type in All Capital Letters. All caps is typically perceived as SHOUTING to the reader and is very difficult to read. Reserve all caps for EMPHASIS.

Always Fill in the Subject Line. Fill in the subject line of the e-mail to alert the reader to the content or purpose of the message. The subject heading should be short, accurate, and informative. In doing so, you give the reader enough information to help them determine the priority of the message as well as how to file the message for future reference in their e-mail filing system. Examples of appropriate subject line headings include Order Confirmation #123456, Confirmation of Appointment on 12/8/05, Status of System Problem, and so forth. Never make the subject line your only message. Type in a message in the message section of the e-mail.

Use the* URGENT *Notation Only in a Truly Urgent Situation. As with the boy who cried wolf too often, overuse of the URGENT notation defeats the point. And, you may think any message you send is urgent; however, your reader may have different priorities. Reserve the URGENT notation only for time-sensitive e-mails that require immediate attention.

Do Not Use Color, Images, or Difficult to Read Fonts. Colored backgrounds or images in the background of an e-mail message make the words very difficult to read. Also, nonstandard fonts are difficult to read. Stick with black and white and basic fonts such as Times Roman, Arial, or Verdana in either 10-point or 12-point type. Avoid bolding or italicizing words unless there is a specific and appropriate reason to do so such as for emphasis or for titles of publications or foreign words.

Note Any Attachments. In some instances, you may need to insert an attachment to your e-mail. If so, notify the reader with in the body message that there is an attachment. It is difficult to identify attachments in some e-mail systems.

Do Not Send Mass E-Mails Inappropriately. Use the CC (copy) or BCC (blind copy) features carefully. Use the CC feature to document that a message has been sent only with those individuals who are immediately involved in the communication. Remember that when using the CC feature, all receivers of the e-mail will see all e-mail addresses. Use the BCC feature when you want someone else to see the message, but you do not want all receivers to see the e-mail address. E-mail addresses should only be sent to people with the address owner's authorization. If you CC others, you have sent out everyone's e-mail address to everyone in the communication.

Do Not Forward Without Permission. In some instances, you may want to forward an e-mail to others to document, inform, or save time. Do not do this without permission from the originator of the message. Although all e-mails should be written with the understanding that anyone can read them, it is common courtesy to get approval before forwarding an e-mail to others.

Begin with a Saluation. As you would do with a customer entering your place of business, begin your e-mail with a warm salutation. Example: *Good Morning, Mr. Johnson* or *Dear, Ms. Smith.* If you are e-mailing a larger group, consider using a salutation such as *Hi Everyone* or *Greetings from the Service Department.*

The Middle

The middle of your e-mail message is found in the message part of the e-mail and communicates both the message and the tone of the message to your customer. Think carefully about what you want to say and how you want to say it and consider the following tips.

Keep Your Message Short. A very long paragraph with no breaks encourages the reader to skim the paragraph, potentially missing important details. Create short paragraphs and list like items in bullets. Double space between paragraphs. Also, use informative headings such as Problem, Solution, Action Taken, Action Requested, and so forth to ensure that the reader focuses on the important points and action items.

Write Your Message as You Would Say It to the Person Verbally. Think carefully about what you want to say to the reader. Put the most important points first. Be very clear. Use short sentences. Keep your writing active rather than passive. For example, instead of *"An appointment for your system installation was made for you on . . .",* rephrase to say *"I made a system installation appointment for you on . . .".* Avoid emoticons or Internet lingo such as LOL (laugh out loud), ROTFL (rolling on the floor laughing), BFN (bye for now), :) (smiling), and so forth. Reserve these shortcuts for your friends outside of the workplace. Also, do not use multiple exclamation points or question marks. These kinds of emphases tend to sound condescending or sarcastic online.

Maintain Your Professionalism. Never forward jokes, personal images, unrelated Internet links, chain letters, or any other unimportant e-mails to customers. (In fact, do not forward these types of e-mails to your friends without their permission.) These e-mails run the risk of containing viruses or spyware, filling up their mailbox or slow down their connection speed, or offending someone you may not know well.

Never Respond to an Angry E-mail Immediately. Take a breath and a break for a few minutes before responding to an angry or rude e-mail. Remember that with e-mail (1) everything you write is documented permanently, and (2) communications often sound harsh anyway. If you write a response when you are angry or defensive, the tone of the message will clearly indicate this. As with all customer relations, it is important to respond calmly and pleasantly. Think carefully about how you respond, read your message over several times, and even ask someone else to read it to ensure that you are not sending the wrong message.

Always Check Your E-Mail for Errors. Before hitting the send button, check your message for (1) the accurate e-mail address, (2) correct grammar and spelling, and (3) the appropriate tone that reflects a friendly and helpful attitude. Remember that with only text to make the impression, you must carefully choose your words. In all written communications with customers, use correct grammar and punctuation. Poorly written sentences and misspelled words send an unprofessional image. The customer may also think that if you cannot write correctly, you cannot serve them correctly in any other way and may take their business elsewhere.

The Ending

As with face-to-face interactions and telephone calls, the end of the e-mail message helps to leave a positive and friendly impression with your customer. Thoughtfully close your e-mail message using the following tips.

Protect Your Computer. Ensure that your computer system is protected from viruses and other damaging programs. You would not want to send your customers a virus from your computer. To ensure that you protect your customers, be sure to maintain appropriate virus protection on your own system. If you send e-mail from your company's computer, it is probably already protected. If you send e-mails from you own computer, be sure to have a security program installed.

Close Your Message Politely and with Purpose. Close your e-mail message with *Sincerely, Thank you, Have a Good Day,* or some other courtesy. Consider adding a sentence that shows how much you care about your customer or how much you appreciate their business.

Sign Your Name, Title, and Company Name. Always end your message with your full name, your title, and your company's name as well as a telephone number at which you can be reached. Save the customer time by supplying the information they need if they want to follow up on the e-mail with a call. Be sure to include your company's name. Do not assume that the customer will remember you as the technician or sales person. They will probably better remember the company through which they are making the purchase.

Study/Career Tips

Never use your business e-mail inappropriately. Avoid the following practices:

- Personal use of business e-mail
- Sending e-mails with confidential information
- Speaking poorly of someone
- Sexual harassment
- Illegal activities

Remember that every e-mail can be traced. Never send an e-mail to someone that could be harmful or misunderstood if other people see it.

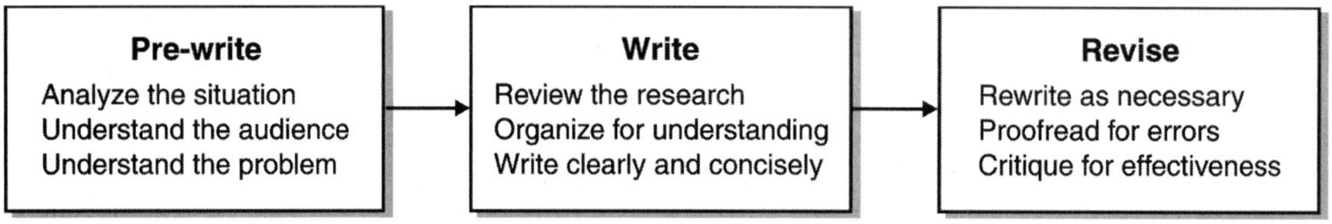

FIGURE 3–4 The Writing Process

Written Communication and Documentation

Much of the day-to-day communication with customers takes place in face-to-face interactions or by telephone, and most professionals are comfortable with these channels. However, written communication is very important and most successful professionals have mastered writing skills. See Figure 3–4 for an effective process to help ensure that documents you write send the correct message to the reader.

Pre-Write

Before you write even a short message, think about the situation. What do you need to say? Why do you need to say it? What impression are you trying to make? What is your goal?

Next, think about the reader of your message. What do they need to know? What is their situation? What are their characteristics that might affect the receiving or understanding of the message?

Finally, review the problem or need. Why are you writing the message? What exactly does your reader need to know or understand? What do you hope to achieve by the communication? Do you have all the information you need to put into the message? If not, conduct the necessary research.

Write

The second step in the writing process is to actually write the message. First, review any research you conducted to solve the problem or answer the question requiring the written communication. Next, organize this information into a cohesive message. Finally, write the message clearly and concisely. It is essential to use proper grammar, spelling, sentence structure, punctuation, and paragraph construction. Poor writing not only leads to a miscommunication of the message, but also leaves an unprofessional, negative impression on the reader. As with face-to-face and telephone communication, the impression you make with written communication is very important and is a significant element in excellent service.

Revise

Finally, before a message is actually sent, it is important to review the message and revise it as necessary. First, reread the entire message and rewrite it as needed. Think about the clarity of the message. Use words that are simple, clear, and unambiguous. Carefully think about how you form your sentences. Will the reader understand what you are trying to convey? Is the message written as concisely as possible? Wordiness often leads to confusion. Look for errors in punctuation, grammar, and spelling. If using a word processor, use the program's spellcheck and grammar check features. If writing the message by hand, use a dictionary if needed. Finally, reread the message in its entirety and make sure you have composed a message that meets the initial need. Remember that your goals of writing are to say what you mean to say and to make a positive, professional impression.

Study/Career Tips

If you have trouble with writing in areas such as grammar, sentence structure, punctuation, spelling, or forming paragraphs, seek help. The ability to write well is very important in excellent service and in your career success in general. There are many ways to improve. Take or review your college courses in English. Seek help on the Internet by searching for writing tutorials. Purchase a basic writing book and study it. Ask others who write well to critique your writing. Use the spelling and grammar tools on your word processor. Also be sure that you handwriting is legible. Anyone can write well with practice and effort.

Common Business Formats

There are several document formats commonly used in business. It is important to understand the basic structure of each format so that you can use the appropriate format for the appropriate situation.

The Memo. The memo is used for relatively short communications that inform, supply information, ask for information, confirm, document, respond, and give directions. Use the following tips to ensure a well-constructed memo:

- **Complete the To, From, and Date headings:** It is important to fill in the appropriate information for these headings to ensure that the correct person receives the message and that the receiver knows who the communication is from. The date helps in the documentation for the future.
- **Complete the Subject heading:** The subject heading states briefly what the communication is about. It should state the subject of the message as clearly and concisely as possible. The subject is often used to find the message at a later date when documentation is needed. Avoid meaningless subject lines, such as *Hello* or *Important* or *Response*.
- **Use a conversational tone:** Typically, memos are used when people are familiar with each other, as in an office or company environment, or with customers with whom you have conducted business for a while. A conversational tone helps to maintain the relationship and is often easier to understand.

- **Be direct and concise:** Memos are typically designed for short and to-the-point communications. Avoid wordiness or lengthy introductions or summaries.
- **Focus on a single topic:** Try to focus the memo on only one topic at a time. If more than one topic must be addressed, use a numbering or outline system to ensure that the memo is organized for the reader.
- **Use bullets, highlighting, tables, and other visual elements:** To help the reader see important points or stay organized, use bullets, highlighting, tables, or other visual formatting or elements as appropriate.
- **Be courteous:** As with any other kind of communication, be courteous to the reader. Reread the message to ensure that you have conveyed the appropriate attitude. Even in a problem situation, courtesy is important.

The Letter. Typically letters are used for sending messages to those outside of the company and when a permanent record or documentation of a message is needed. Business letters are typically more formal than memos or e-mails. Letters also allow for more confidentiality than memos or e-mails. Letters should be used for requesting information, providing information, documenting activities or decisions, replying to questions or requests, telling bad news or denying requests, and persuading the reader to take action or to buy a product or service. The following list represents several strategies that lead to effective letters:

- **Write clearly and concisely:** As with all communications, strive to be clear and concise in the message. Be organized, think about the words and sentence structure you use, and use bullets or numbering when appropriate.
- **Always use a positive and professional tone:** Remember that letters are also opportunities to provide excellent service and to show your customers that you value them and appreciate their business. A positive tone, even in a negative situation, typically results in a better reaction from the reader.
- **Use an opening, a body, and a closing:** As with other forms of communication, there is a beginning, a middle, and an ending to the message. The opening should tell the reader the purpose of the letter. The body should provide the details. If you are providing instructions, be very clear. Use numbering or bullets to help the reader understand the instructions. The closing should provide a concise summary and request a specific action.
- **Arrange information logically:** Think carefully about how you want to organize the information before writing your letter. Arrange information in logical order, such as sequence of events, categories, steps, or order of priority.
- **Open with a positive statement:** Try to always open your letters with a positive statement, even if the purpose of the letter is negative. You might thank the customer for their business, state that the order has been received, state that you appreciate the customer's concern, and so forth.

Study/Career Tip

Look up the proper formats for business letters in a writing textbook or on the Internet. Consider keeping a letter and memo template on your computer so that you can fill in the appropriate information without having to review the actual format.

Listen, Listen, Listen!

As discussed in Chapter 2, effective listening skills are essential in both connecting with your customer and communicating with them. A large part of excellent service, problem solving, and even selling to customers is listening effectively.

How well do you currently listen? Rate yourself from 1 to 5 on the following statements where 1 = I need improvement and 5 = I do this well now. Complete Figure 3–5 honestly.

If you rated yourself below 5 on any item, think about focusing on one or two of the strategies at a time until you have perfected your listening skills.

There are several strategies that you can you use to improve your listening skills.

Use Appropriate Body Language

Face the speaker, make eye contact, use good posture, and stay relaxed. Avoid body language that suggests you are angry, closed, distracted, disinterested, or tired.

Use Verbal and Nonverbal Cues

Nod your head in agreement or understanding, lean toward the speaker, encourage the speaker to continue with an appropriate *"uh huh"* or *"I understand."*

Avoid Distractions

Shut out physical distractions such as noises, other people, cell phones, kids or pets, traffic, and work that might be going on around you. If necessary, move to a more suitable area that has fewer physical distractions. Be sure to silence your cell phone or pager. Shut out mental distractions by intently focusing on the speaker and not on what you are going to say next, what you assume to be the problem or solution, or what you plan do next in the day.

Ask Questions

Once the speaker has finished the thought, ask appropriate questions that (1) show you have listened attentively, and (2) clarify the exact meaning of the message. Continue to ask questions until you are certain you understand what the speaker is intending to say. Also consider asking questions in your own words or paraphrasing what the speaker said.

	Rating
I make good eye contact with the person speaking.	
I keep my mind focused and avoid thinking about anything else except for what the speaker is saying.	
I am active and responsive when I listen and use verbal and nonverbal cues to show that I am listening.	
I wait until the speaker has completed the thought before starting to talk.	
I keep an open mind until all of the information has been discussed.	
I focus on the content of what is being said rather than being destracted by the speaker's mannerisms or appearance.	
I avoid distractions around me as I listen to the speaker.	
I listen for main ideas and the most important points the speaker is trying to get across.	
I ask questions to ensure that I fully understand what the speaker is trying to say.	
I summarize or paraphrase what the speaker has said to ensure that I fully understand the intended message.	
I leave my emotions out of the conversation so that I can fully understand the speaker's message.	
I avoid arguing mentally while the speaker is talking.	
I don't anticipate what the speaker is going to say before it is said.	
I ignore or avoid jumping to conclusions or making assumptions.	
I empathize with the speaker and try to place myself in his shoes.	
I pay attention to the speaker's nonverbal communication, such as facial expressions, gestures, postures, etc.	
I take notes, if needed.	

FIGURE 3–5 Listening Skills

Don't Interrupt

Don't interrupt the speaker until she has finished the train of thought, even to clarify meaning. Allow the speaker to finish. Finish listening and digesting the information, then speak yourself.

Be Organized in Your Listening

Listen for the main points. Take notes if the information is detailed or technical. Form mental pictures of what the speaker is saying. Pay attention to phrases such as "*My point is . . .*" or "*My problem really is . . .*"

or other obvious statements of importance. Also pay attention to those details that are repeated often.

Keep an Open Mind

Listen thoroughly before you try to solve the problem or make a decision. Get all the facts first. Analyze your assumptions and try to eliminate them.

Pay Attention to the Speaker's Nonverbal Cues

Pay attention to what is not said as well as what is said. Is the speaker angry, emotional, or frustrated? What does the speaker's body language say? Watch gestures, posture, and other nonverbal cues to try to understand the entire message.

Analyze and Minimize Your Own Barriers to Listening

Assess if there is a language barrier, bias, prejudice, internal stress or insecurity, preoccupation, or any other barrier that might hinder your effective listening. Sometimes just being aware of a barrier to listening reduces its impact.

Questioning for Understanding

Learning how to ask questions to ensure that you understand your customer serves two important objectives: (1) ensuring that you actually understand the customer so that you can complete the task, solve the problem, or provide information effectively and accurately, and (2) ensuring that the customer feels that you value him, that you hear him, and that you are on his side. This second objective is important even when you cannot provide the service or solve the problem. In many cases, you have gained a loyal customer for future business because you have listened carefully and taken the necessary actions to understand.

Asking the right question is not always as easy as it might seem. There are many strategies that can make questions effective. Let's review a few of these questioning strategies.

Ask Open-Ended Questions Appropriately

Open-ended questions are those questions that seek more than a one- or two-word response, especially yes and no. The goal of open-ended questions is to gain a large amount of information. They ask the customer to think about the issue or need. They also ask for feelings and opinions. Examples of open-ended questions include:

What can I do for you?

What would you like to know about this product?

Where do you think the problem lies?

What do you mean by . . . ?

How did you first discover this problem?

How do you intend on using this service most of the time?

Each of these questions asks the customer to explain thoroughly what they want or need.

Advantages of open-ended questions:

- They tend to build trust and set up a situation in which you and the customer are on the same side.
- They tend to be less threatening.
- They allow a free response rather than a guided response based on potentially inaccurate assumptions.
- They tend to result in a large amount of information.
- They can be used to solicit many different ideas and opinions.

Disadvantages of open-ended questions:

- They are time-consuming.
- They can result in unnecessary information.
- They require that the answerer understand more of the issue.
- The customer is in control of the conversation.

Good uses of open-ended questions include:

- Use open-ended questions when you want to set the customer at ease and ensure that they feel heard.
- Use open-ended questions with very quiet or reserved customers to encourage them to converse.
- Use open-ended questions to get customers to see the extent of their problems so that they understand that it may not be as simple to fix as they first thought.
- Use open-ended questions to develop rapport with customers by asking about their kids, their new home, their neighborhood, and so forth. These questions are good relationship-building tools.
- Use open-ended questions to find out as much information as possible about the customer's wants, needs, expectations, problems, concerns, future needs, etc. The more information you have, the better you can serve your customer.
- Use open-ended questions when you are brainstorming for problem solutions with a coworker or contractor. Example: *What could be done to . . . ? How could we design this so that . . . ? What solution would you suggest for . . . ? What do you think would happen if we . . . ?*
- Use open-ended questions to learn new information from coworkers or contractors. Example: *What do you think about . . . ? Why do you say . . . ? How would you decide about . . . ? Why did you install it this way?*

Ask Closed-Ended Questions Appropriately

Closed-ended questions are those questions that have a definite answer such as yes or no or some other factual response. The goals of closed-ended questions can be to gain facts, to probe for more information, or to direct the responses. Examples of closed-ended questions include:

Do you understand how to use this system?

Do you know what the problem is?

When did the problem start?

Do you want the system installed in this room?

Is this what you are looking for?

Does this answer your question?

Each of these questions asks for specific information that can be answered in a word or phrase.

Advantages of closed-ended questions:

- They are quick and require only the answer without a lot of discussion.
- The response provides only the information requested.
- They are easy to answer.
- They allow you to keep control of the conversation.

Disadvantages of closed-ended questions:

- The responses often provide incomplete information.
- The questions can be leading.
- The questions are dependent upon the questioner's knowledge and assumptions.
- They tend to be more threatening and irritating to customers.
- They discourage the customer from providing all of the information.

Good uses of closed-ended questions include:

- Use closed-ended questions to start off a conversation with a customer you do not know well to make it easy for the customer to answer.
- Use closed-ended questions to test the customer's knowledge.
- Use closed-ended questions when you want to confirm what you already know.
- Use closed-ended questions to set up a positive or negative frame of mind in your customer. For example, if you want to set up a positive frame of mind, ask several questions that have *yes* as the answer.
- Use a closed-ended question to seek closure to a conversation or sale of a product or service. Example: *Look's like you need the next level of service. Shall I sign you up now?*
- Use closed-ended questions when you want to measure something. Example: *How much did . . . ? How many were there? How often does this happen?*

You can turn any question or phrase into a closed-ended question by adding "*Isn't it?*" or "*Don't you?*" or a similar tag phrase to the end.

Keep Questions Simple and Use Language and Terms that the Customer Can Understand

Remember that for some customers technology can be very foreign and confusing. Ask simple questions and ask questions one at a time. If you need a lot of information, break down the questions into small, simple ones to help the customer maintain his understanding. Try not to sound as though you are interrogating a customer. An interrogation can be avoided by interjecting some open-ended questions periodically with closed-ended questions.

Your customers may not understand technical terms, acronyms, or the common jargon you use in your industry. Even terms such as ISP, display, CPU, USB cable, and so forth may be completely foreign to a customer. Be sure to translate technical terms into common terms unless you are sure your customer understands the technical vocabulary. On the other hand, it is important to not insult your customer by simplifying terms too much. Try to judge the knowledge level of your customer and use terms accordingly. You can tell if they do not understand by a puzzled look, questions, or illogical responses. The goal is to be as clear as possible in light of the customer's knowledge level without being condescending or patronizing.

Ask Probing Questions

The purpose of a **probing question** is to solicit more information from the customer within a specific line of questioning or on a specific topic. You might need to clarify a customer's needs, see what the customer actually expects, determine exactly where the problem lies, or find out what other products and services the customer may be interested in purchasing. Probing questions should be simple and should cover only one point at a time. They not only help you to get the information you need, but they can show the customer that you are interested and are paying attention. You cannot ask a probing question if you have not been listening to the previous answers.

There are several different types of and purposes for probing questions.

- **Clarification questions** can be used when the customer is vague and you need more information. Clarifying questions ask for more facts or description. The customer should not have to think much about answering this kind of question. They should just have to report what they know or have observed. Example: *Can you explain more about the sound you get when you turn off the system?*
- **Evaluation questions** can be used when you need for your customer to offer an expert opinion, such as working with a contractor. Example: *What are the pros and cons of installing this system in this manner?*
- **Example questions** can be used when you do not understand the customer's explanation. By asking for examples, you receive concrete descriptions or explanations. Example: *I'm sorry, but I don't really understand. Can you give me an example of when this system does what you are saying?*

- **Purpose questions** can be used when you need to understand the underlying reasons for something, such as when you are working with a contractor and are asked to complete a task in a very specific way. Understanding the purpose might help clarify what they need. Example: *So, what are you thinking about the problem here and how will this fix resolve it?*
- **Extension questions** can be used when the customer has not provided sufficient information and you need more. Example: *Can you tell me more about the situations in which you lose your connection?*

Here are some tips for asking probing questions.

- Make sure that you are listening objectively to the customer and not inserting your own agenda or assumptions into the questions and answers.
- If you are certain of the problem and just need to validate your assumptions, create your probing questions thoughtfully to cover all of your bases.
- Try to be specific with some of your questions to direct the customer to specific areas of a problem.
- Avoid yes or no questions.
- Use probing questions to help establish a relationship with the customer that says "*We are in this together.*"
- Use probing questions to help the customer think in new directions or about different solutions.
- If probing for more personal information, be very careful that you are not insulting or intruding on the customer. Personal probing questions must be used very carefully and tactfully.
- Never ask questions that insult, attack, irritate, or make the customer feel stupid.

Summary

In Chapter 3, we discussed important aspects associated with effective communication with the goal of providing excellent service. First, the communication process was explained as a tool for any type of communication. This process can be used to help ensure the intended message receiver actually receives the sender's message and understands it accurately. The process can be used to identify the potential barriers for effective communication. Next, several channels of communication commonly used in the business world were identified and discussed including face-to-face communications, the telephone, e-mail, and communications using written documents.

In addition to the communication process and types of communications, several important techniques for ensuring effective communication were discussed. The important skill of listening was explored, identifying many strategies to assist communicators in listening effectively. Additionally, the important skill of asking the appropriate question was explored. Many different types of questions were described.

Important points in this chapter are:

- The communication process describes a step-by-step procedure that can be used to ensure that important elements of any type of communication are addressed.
- There are several channels of communication that can be used for different purposes: face-to-face communications, the telephone, e-mail, and written documents.

- Regardless of the type or channel of the communication, a sender of a message makes an impression on the receiver. A positive impression is important in excellent service to customers.
- Every communication, regardless of the channel, has a beginning, a middle, and an ending. There are several strategies that can be used to ensure that the message is communicated as intended and achieves the desired goal.
- There are several specific strategies that can be followed to ensure that e-mail and written communications are received as intended and make the desired impression.
- Effective listening skills are essential for effective communication. There are several important strategies that should be made habits.
- There are numerous kinds of questions that can be used to achieve communication goals with customers. It is important to understand the different kinds of questions and how they can be used appropriately.

Key Terms

Clarifying question A clarifying question is one that asks for more facts or description.

Closed-ended question A closed-ended question is one that asks for a definite answer such as yes or no or some other factual response.

Communication channel The way in which a communication is sent or the type of media that is used such as face-to-face, computer, e-mail, written document, telephone, etc.

Communication process The communication process outlines four steps that should be followed for effective communication. The process includes (1) encoding and sending of the message, (2) transmission of the message through an appropriate channel, (3) receiving and decoding of the message by the receiver, and (4) interpretation and understanding of the message by the receiver.

Evaluation question An evaluation question is one that can be used when you need your customer to offer their expert opinion.

Example question An example question is one that can be used when you do not understand the customer's explanation.

Extension question An extension question is one that can be used when the customer has not provided sufficient information and you need more.

Netiquette Netiquette represents the manners or professional courtesies followed in an on-line communication.

Open-ended question An open-ended question is one that seeks more than a one or two word response.

Probing question A probing question is one that solicits more information from the customer within a specific line of questioning or on a specific topic.

Purpose question A purpose question is one that can be used when you need to understand the underlying reasons for something.

Review Questions

1. What are the four major steps in the communication process?
2. Why is it so important to make a great first impression in face-to-face communication with customers?
3. What are six different strategies you can use to make a face-to-face communication with your customers positive?
4. What are seven different strategies you can use to make your telephone communication with your customers positive?
5. Why is it important to place your customer on hold using the hold feature of the phone if you must be gone from the phone for a long period of time? What are other options to placing your customer on hold?

6. Assume that you must wait for information on your computer while you work with your customer on the telephone. What communication strategies should you use with the customer in this situation?
7. Assume that you are taking a phone message for a coworker who is late to work because he overslept. What do you tell your customer?
8. How should you end a phone call with a customer? List four specific ways to bring closure to the call.
9. What are three considerations for making a good impression when you call a customer on the phone?
10. What information should you put in your voicemail message?
11. When leaving a message on your business voicemail, what information should you provide for your customers?
12. What are ten different practices that you should follow when e-mailing a customer?
13. How might a customer interpret your e-mail when you type in all capital letters?
14. What are eight different strategies for effective listening?
15. What are potential barriers to listening to your customers?
16. What are advantages and disadvantages of an open-ended question? What are situations in which this type of question should be used?
17. What are advantages and disadvantages of a closed-ended question? What are situations in which this type of question should be used?
18. What are five different types of questions that you can use with a customer to get information, and when should each type be used?

Handling Difficult Customer Situations

OBJECTIVES

After studying this chapter, you should be able to:

- Explain strategies for handling dissatisfied customers.
- Explain strategies for handling difficult customers.
- Explain effective troubleshooting and problems-solving skills that can be used with customers.

OUTLINE

Introduction

Handling Difficult Customer Situations

Troubleshooting and Problem Solving

Conclusions for Difficult Customers and Customer Situations

Introduction

Though the goal of all interactions with customers is to assure that they are satisfied with the products and services, things will not always go as planned and customers will no doubt sometimes have problems, be dissatisfied, and be difficult to deal with. In these cases, excellent service is critical. The excellent service provider can use many different strategies and techniques to ensure that the problem is solved, that the customer is ultimately satisfied, and that even difficult customers are treated with respect and courtesy.

The idea that the customer is always right is probably not very true in reality. Customers can be unreasonably demanding. They can be rude and discourteous. They can often cause their own problems. However, even these customers are just that—customers. They are paying for products and services and deserve excellence. They have the potential to be customers in the future, sustaining the business.

And, sometimes, the most disgruntled customers can be transformed into the most loyal customers if treated well.

This chapter discusses several common problem-customer situations and provides several techniques and strategies that may be helpful in providing excellent service in difficult situations.

Handling Difficult Customer Situations

Customers can be difficult to deal with for many reasons. Some individuals are just unpleasant and tend to spread their unhappiness to everyone in their path. Some customers may be great people, but may be having a bad day and are taking their problems out on others. Other customers are self-focused and demand perfection and even services and products that are unrealistic. Some customers may be legitimately upset because of a failed product, poor service, a mistake, a mishandling of a situation, or some other situation.

How you handle difficult customers is very important in meeting their needs, making them feel good about doing business with you, developing and maintaining customer loyalty, and in ensuring future business. In short, how you handle difficult customers is one of the most important elements of excellent service. Let's explore some common difficult-customer situations and look at how to apply the problem-solving process and effective communication skills to provide excellent service.

There are many common problem situations in which you may be faced with dealing with a difficult customer. Fortunately, there are also many effective strategies that can help. Study each strategy for the following situations carefully. You can then adapt or combine appropriate strategies to meet your unique situation. Remember that the objective is to provide the best service possible with the goal of maintaining your customer loyalty and to keep your customers for the long term.

Critical Thinking Question

- What are potential customer problems you think you may face in your job?

List as many different kinds of problems as you can think of.

Schedule- and Delay-Related Problems

It is important to be on time and to stay on schedule with your customers when possible; however, there may be times when staying on an agreed upon schedule just is not possible. There may be situations beyond your control or you may find that a task takes longer than you estimated.

The first strategy is in prevention. Learn to accurately estimate how long it takes to go from one place to another, how traffic typically causes delays at specific times of the day, and how long it takes to do common tasks. Plan for problems. You can do this by keeping a map in your car, learning alternate routes that take you around traffic congestion, ensuring you have all the equipment and parts needed for a

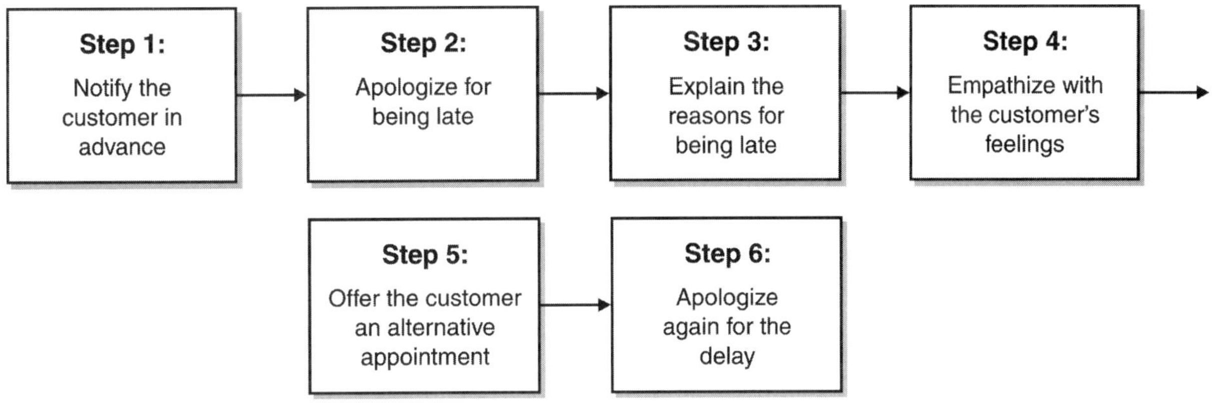

FIGURE 4–1 Steps for Schedule or Delay Related Problems

job before you make the appointment, and so forth. Make it a habit of walking through an entire project from beginning to end, including travel time and equipment ordering, and then give yourself a little breathing room for any problems that come up.

If you are running late, use the following strategy to minimize the inconvenience to the customer and keep the customer as satisfied as possible. These steps are summarized in Figure 4–1.

Step 1: Notify the Customer in Advance

If at all possible, tell your customer that you will be late as soon as you know. Call ahead and give the customer as much advance warning as possible. Even if you are going to be delayed only a few minutes, it is important to tell this to the customer in advance. It shows that you are very concerned about being on time and about your customer's time. Ten minutes late is still late and, although it may not matter in the long run, it makes a negative impression.

Step 2: Apologize for Being Late

Take personal responsibility and do not make excuses or blame others. People find it difficult to be angry with someone who is sincerely apologetic and who takes responsibility for actions.

Step 3: Explain the Reasons for Being Late

Legitimate reasons are not the same as flimsy excuses. Excuses are used to get forgiveness from the customer. Reasons are simple facts that, if you have planned well, are simply beyond your control. Most customers will understand those things beyond your control if they have seen that you did your part in planning. If your reasons are not legitimate, such as forgetting an appointment or over sleeping, it may be best to give a more general statement such as *"I was unavoidably detained."*

Step 4: Empathize with Your Customer's Feelings

Delays obviously can be inconvenient. It is easy to become defensive when your customer does not seem to care that you planned well, but that the delay is beyond your control. Rather than becoming defensive, empathize with your customer's feelings of frustration. Tell them you understand that you have caused an inconvenience.

Step 5: Offer the Customer an Alternative Appointment

Offer to reschedule for a more convenient time or provide some other alternative to give the customer a feeling of control over the situation. There is nothing more frustrating than being at the mercy of someone else. Allowing the customer to maintain control over the situation often reduces the frustration. You might also be able to suggest a way that you can make up for the delay so that the task is completed on time even with a late start. Ideas might include asking another worker to help, rushing an equipment order, paying for express mail, etc. Look at the situation and be creative in solving the problem you may have created by being late.

Step 6: Apologize Again for the Delay

Once the task has been completed successfully, apologize again for the delay and thank the customer for his patience and business. Customers appreciate a sincere apology. An insincere apology usually makes the problem worse.

You can adapt these steps when a customer is in a hurry and wants you to complete your task quickly. Accurately estimate how long the task will take. Explain reasons that you cannot complete the task faster, empathize with the customer's urgency, and offer alternatives or creative solutions. In this case, be sure to echo your customer's urgency in your tone of voice, in your actions, and in your demeanor. Let the customer know that you care about the need for urgency and doing what you can to meet the need.

Critical Thinking Questions

- What are the consequences of not dealing with schedule- and delay-related problems appropriately?
- What negative impact would these kinds of problems have on your customer and your company as a whole?

Study/Career Tip

Consider carrying a calendar with you as you work to take down appointment information and other customer notes. An alternative is a laptop computer with a calendar or notebook program. It is important to ensure that you do not miss or mix up appointments with customers. Even if your company provides you with this information, a notebook or calendar provides a way to write down customer questions or notes on customer problems and other information you need to provide excellent service.

Customer Complaints

Effectively dealing with customer complaints is essential for many reasons. First, welcome customer complaints. Most customers walk away from you (usually to your competition) without saying a word and without giving you the opportunity to correct the problem. When this happens, not only have you lost the customer, but you also do not have a chance to fix problems for future customers. You have lost future business that this dissatisfied customer might have referred to you. Remember that disgruntled customers tell their friends about their problems and poor treatment at every opportunity. However, a customer with a problem that has been successfully resolved with great service also spreads the word. She will quickly tell others how well she was treated. Also realize that customer complaints are typically genuine. It takes a lot of time and effort to complain. Most people will not do this unless there is a legitimate problem that they need to have addressed. Customer complaints deserve your full and prompt attention.

Apply the problem solving process to a customer complaint to solve the issue as quickly and efficiently as you can. Let's review how the process can be applied to a customer complaint. As you can see, some of the steps have been combined. Figure 4–2 provides a summary of the steps.

Step 1: Research the Customer's Problem

Researching the problem starts with the customer. Take sufficient time to really listen to what the customer is saying. Take notes if you need to about important details. Use all of the effective listening skills to make sure that your customer feels that you are listening and that you care about the problem. Remember that listening serves two purposes. First, it allows you to research the problem carefully from the customer's perspective. Second, it tells your customer that you are concerned, that you plan on doing something about the problem, and that you value them. Verify that you understand everything you need to understand to successfully resolve the problem. Verification is also accomplished with effective listening skills and appropriate questioning (Chapter 3). You may also need to do some additional research to determine the cause of the problem and to identify appropriate solutions.

Step 2: Implement the Best Solution to Fix the Problem

Evaluate each solution that came up in the research process and implement the best solution. Assess how well the solution works by discussing

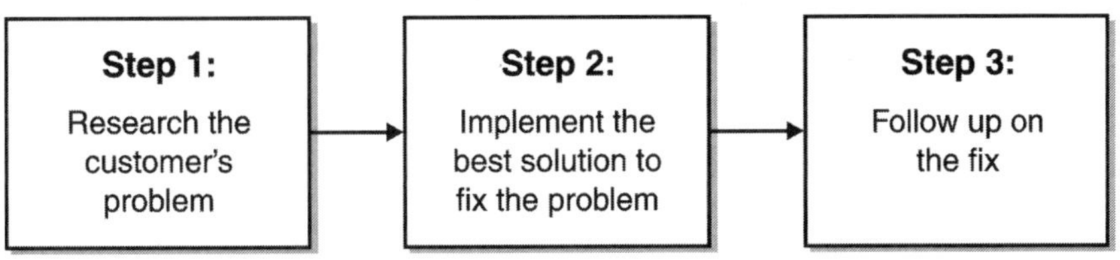

FIGURE 4–2 Steps for Dealing with a Customer Complaint

it with the customer or testing the equipment. If discussing the solution, go over each step that you took and ask the customer if the solution works for them. Also tell the customer how you will prevent this from happening to other customers in the future, if this is appropriate. Fixing the problem for future customers is critical. Remember that most customers do not complain. They just walk away. If you do not fix the problem for future customers, it is guaranteed that you will lose customers because of this same issue. By telling your current customer how you will prevent the problem in the future, you demonstrate that you care about customers in general. The current customer will feel good that they did make the complaint and helped future customers.

Step 3: Follow Up on the Fix

Even if you know that you have fixed the customer's problem, follow up shortly after the occurrence. Call the customer to see if the problem is still fixed and if there is anything else you can do for them. This simple gesture will go a long way to building a long-term relationship with your customers. They will appreciate your concern and feel valued. Thank them for their continuing business and leave your contact information in case they have additional questions.

Employee Mistakes

You can use similar strategies to address the situation where you or a coworker has made a mistake with the customer. Mistakes do happen. In some cases, someone else in your company or a partnering company has made a mistake and has left you to deal with the customer. These situations can be very difficult, depending upon who made the mistake and the cost of correction. In many cases, a future relationship may be at risk either between your company and the customer or your company and a partnering company. Many mistakes are the result of poor communication and documentation with the contractor, designer, dealer, or customer. Documentation is critical. All aspects of a project should be reviewed carefully before even beginning the project.

Here are some basic truths that apply to any mistake:

- Bad news never improves with age. Strive to resolve problems immediately.
- Problems are easier and cheaper to fix before they go too far downstream.
- If you want to do business with that builder or electrician again, do not discuss their mistakes with the customer in their absence.
- It is best to resolve a problem with the builder or tradesman in private and out of sight of the customer.
- Sometimes it is ultimately cheaper and simpler to just fix it and move on.
- Time costs money, including the time you spend blaming someone else.

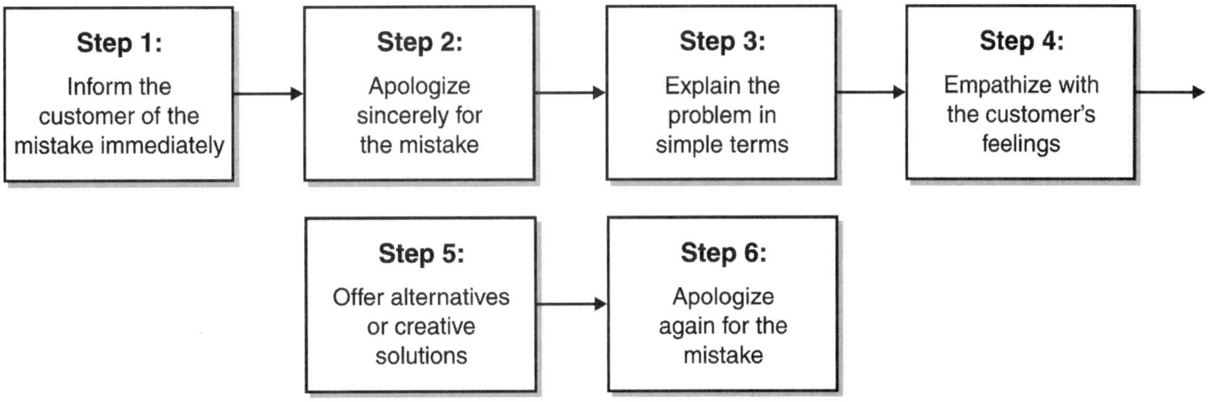

FIGURE 4–3 Steps When You Have Made a Mistake

In these difficult mistake-related situations, use the following adaptation of the problem-solving process to remedy the situation and maintain customer satisfaction and loyalty. Figure 4–3 outlines some basic steps to help in dealing with this kind of customer problem.

Step 1: Inform the Customer of the Mistake Immediately

Inform the customer of the mistake as soon as you can. Hiding a mistake or waiting for the customer to discover it usually results in even more problems. It is typically best to address mistakes as soon as possible and to discuss serious mistakes in person.

Step 2: Apologize Sincerely for the Mistake

As with being late, sincerely apologize to the customer for the mistake and take responsibility without giving excuses. Remember that you are representing not just yourself, but the company as a whole. If another person made the mistake, apologize on behalf of the company. Customers will often respect individuals and companies who take responsibility.

Step 3: Explain the Problem in Simple Terms

Explain the cause of the problem or how the mistake occurred in terms your customer can understand. Do not give excuses for mistakes, but address them as what they are—mistakes. It is common to place the blame on someone who is not present or even to blame the customer in some cases. Avoid doing this. State the facts clearly and take responsibility for the situation.

Step 4: Empathize with the Customer's Feelings

Try to predict what your customer feels and offer empathy. Show that you understand how the mistake is causing frustration or inconvenience or lack of faith in the competence of the company. Listen to the customer's concerns or simply allow them to vent. Being defensive never helps the problem, even when there are legitimate defenses.

Step 5: Offer Alternatives or Creative Solutions

If possible, develop an alternative or solution that will fix the resulting problem even before notifying the customer of the mistake. Remember to consider inconveniences in terms of time or schedule changes, the value of the customer's time, additional costs, or other problems related to the mistake. Take responsibility for any additional costs incurred. If possible, make up for the mistake with discounted costs or upgraded products or services, depending upon the situation. Remember that your ultimate goal is to maintain a loyal and satisfied customer. Spending a few dollars may save or earn your company many dollars over the long term. Be aware that how you deal with mistakes both personally and as a business is an important factor in providing excellent service.

Step 6: Apologize Again for the Mistake

Once the problem has been resolved, apologize again for the mistake, review how the problem was solved, and thank the customer for her understanding and business. Consider following up with a call or brief note shortly after the incident reiterating the apology and how much you appreciate the customer's business.

Critical Thinking Questions

- When an employee blames others or makes excuses for mistakes, what impression does this make on the customer?
- What might your customer think?

When Equipment Fails

When dealing with equipment, equipment failure is always a possibility. This situation should be treated using the problem-solving process. In most circumstances, you should immediately accept responsibility. The customer hired you because you are the expert. If the equipment you recommended fails, you need to fix it or replace it. You should do so with a pleasant attitude and a sense of urgency for replacing it or getting it fixed.

Note: There are two situations in which you probably will not be able to fix or replace the equipment: (1) when the customer provided their own equipment, and (2) when a customer insisted that you use a specific brand or type of equipment that you opposed or were unfamiliar with. In both situations, it is important to advise your customer in writing about the risks and the limits of your support before the contract is signed and the work begins.

Negative Customer Attitude about Your Company

Customers who have spent large amounts of money with your company and have been disappointed in the products or services will no doubt develop a negative attitude about future products and services and

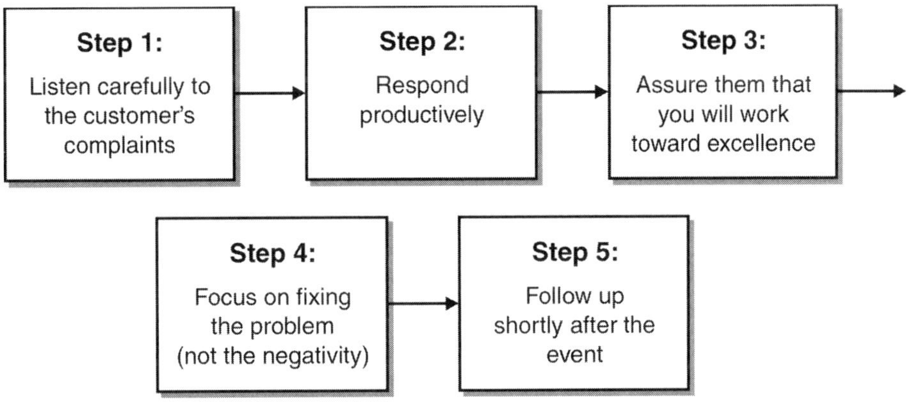

FIGURE 4–4 Steps for Dealing with Customer Negativity

about your company in general. In many cases, he cannot go elsewhere because of the initial investment or limited options. In some cases, your products may rely on the products and services of another company, such as a phone, cable, or security company that has caused difficulties for your customer in the past. In other cases, your customer may have heard negative things from others about your products or services and is predicting that he will be treated in the same fashion. In these cases, consider it a challenge to regain his faith and satisfaction. Figure 4–4 outlines steps that may help in dealing with this kind of problem.

Step 1: Listen Carefully to the Customer's Complaints

Sometimes customers just need to vent about the past or about what they have heard from others. It may be that no one else has acknowledged the previous problems. Listen actively to the customer and be empathetic. Use effective listening skills. For example, one manager experienced a situation with a customer where there was an issue with a very large project with several hundred thousand dollars at risk. The customer had received an invoice for a few hundred dollars that he did not understand from a contractor who happened to be an engineer. The first thing the manager did was to send both parties a humorous article about engineers to lighten things up. The manager was then able to resolve this minor issue that had grown like a cancer and free up the bottleneck that was hindering the progress of the installation project. This issue was not about money, it was about the customer's perception that no one was listening.

Step 2: Respond Productively

As you respond to the customer's complaints, do not get defensive or let the customer lure you into a debate. Never argue, try to change their opinion, or provide excuses for previous poor service. On the other hand, do not encourage the customer to carry on unreasonably or to continue to focus on the negative attitude. Simply let them know that you are listening and that you care. Take on the situation enthusiastically as a challenge to regaining the customer's satisfaction and loyalty.

There are several tricks you can use to help reduce the negativity and to remain neutral in the discussion. Neutrality is important to successfully solving the problem. Remember that you do not want to be defensive; however, you do represent your company and sometimes your company's partners.

Indirectly Acknowledge the Negative. By using an indirect approach, you do not have to directly agree with the customer, which might put you at odds with your company or feed your customer's anger. An indirect approach does allow you, however, to acknowledge the customer's feelings. Use a response such as *"Some people do think . . ."* or *"I am sure that others have experienced something similar . . ."*. By validating the customer's feelings and experience, you may reduce their frustration and anger. By using indirect phrases, you avoid agreeing with your customer that your company is bad.

Rephrase Using Less Emotional Words. Words like "mistake," "problem," "angry," and "lie," tend to evoke negative emotion. Try to rephrase the customer's statements and replace these types of words with those that evoke less emotion, such as "misunderstanding," "challenge," "miscommunication," and "misinterpretation."

Avoid Absolutes. Words such as "never," "always," "only," etc., tend to increase emotion and negativity with an angry customer. Look at the following examples: *"Our installers never make these kinds of mistakes." "The only option you have is to reorder the part." "The scheduler always confirms an appointment 24 hours in advance."* If you were an angry customer, how would these kinds of statements make you feel? Now replace these statements with the following: *"It is unlikely that our installer made this kind of mistake." "Perhaps we could reorder* the *part to get this fixed as quickly as possible." "Our schedulers typically confirm appointments 24 hours in advance."* Softer words go a long way toward reducing the customer's frustration and tend to avoid putting the customer on the defensive.

Study/Career Tip

What are other words or phrases that tend to evoke negative emotions with people? Make a list of as many negative words and phrases as you can. Try to develop the habit of avoiding these words in your conversations with customers. Many aspects of excellent service are habits that simply need to be developed.

Step 3: Assure them that You Will Work Toward Excellence

Tell the customer that you understand that they have had a negative experience in the past, but that you will do your best to see that they now have a great experience. Assure the customer that you will make every effort to fix the problem and regain their trust. Take personal responsibility. Show that you appreciate their business.

Mr. Allen, I am going to change your view of the ABC Company by fixing your problem.

Mrs. Jackson, I am going to make it my goal to turn you into a very satisfied customer. Let's see how we can fix this issue.

Step 4: Focus on Fixing the Problem (Not the Negativity)

Continue to refocus the customer on the problem at hand and how you can help to fix it. Disgruntled individuals tend to continue to complain about the past. Since you cannot change the past, your task is to fix the current problem and attend to the customer's immediate needs. Allow the customer to vent only to a point and then gently refocus to the current solution.

Step 5: Follow Up Shortly after the Event

Shortly after the event, follow up with the customer. Be sure that the problem has been resolved. Thank the customer for his patience and for the opportunity to make things right.

Not Knowing the Answer

There will inevitably be a situation where you just do not know the answer or what to do. Your options are to tell the customer that you do not know or to "fake it" and hope for the best. Obviously, the best option is to tell the customer that you do not have the answer. Remember that it is your responsibility in providing excellent service to ensure that the customer's needs are met. Passing the problem on to someone else is not a very good option. Even if you must make a referral to another party, you are still responsible for seeing that the customer is satisfied in the end. Use the following adaptation of the problem solving process to achieve the best results and to gain your customer's respect. Figure 4–5 illustrates these steps.

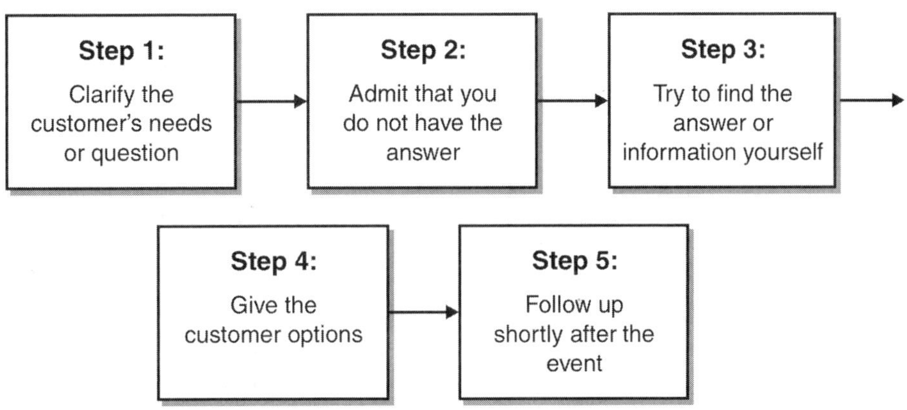

FIGURE 4–5 Steps When You Do Not Know the Answer

Step 1: Clarify the Customer's Needs or Question

Use effective listening strategies to ensure that you understand exactly what the customer needs or that you know exactly what the customer is asking. Rephrase the customer's statements and ask appropriate questions.

Step 2: Admit that You Do Not Have the Answer

Rather than being evasive or providing inaccurate information, simply admit that you do not know the answer or cannot fix the problem. Most people will respect this response. Any other response could result in an even more serious problem or in your customer losing faith in you.

Step 3: Try to Find the Answer or Information Yourself

In some cases, you need only to conduct some fast research to find the answer. If this is possible, find the information yourself. In other cases, you may have to go to someone else for the information. Explain to the customer what you are going to do and how long it will take. When people understand the tasks that you must complete, they are more likely to give you the time you need to complete those tasks.

Step 4: Give the Customer Options

If it is going to take a while to find the answer, or if you need to consult with someone else to get the information, inform the customer and ask what he wants to do. He may want to hold (if on the phone) while you make a call, or you may need to call the customer back. If you are to get back with him, let him know when to expect the call or when you will return with the information. If the best option is to refer the customer to someone else, be sure to facilitate this interaction by (1) providing the information you have gained in your research or an idea of the cause of the problem to the customer, (2) informing the new contact to expect a call from the customer, or (3) asking the new contact to call the customer. If you are referring a customer to a third party, it is important to know if the new contact can actually help the customer. It is a good idea to contact the third party first. Complete as much of the work for the customer as you can. Try to eliminate the need for the customer to repeat the description of the problem to others.

Step 5: Follow Up Shortly After the Event

If you have referred a customer to someone else, be sure to follow up to make sure that the customer is satisfied. Otherwise, you run the risk of giving the customer the "run around."

Inappropriate Actions by Customers

Unfortunately, many customers are uncooperative or just plain rude. They can yell at employees, use inappropriate language, be uncooperative in the steps needed to fix the problem, and so forth. In some cases,

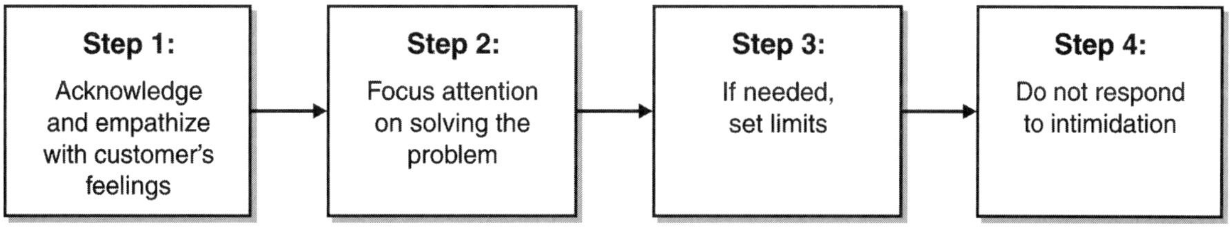

FIGURE 4–6 Steps for Dealing with Inappropriate Customers

customers can actually become abusive or threatening. There are several strategies you can use to effectively deal with these customers while maintaining the integrity of your company and protecting yourself and those around you. The first goal is to stop the inappropriate behavior. Once the customer is behaving appropriately, you can then take the necessary steps to address the initial problem. Figure 4–6 outlines some useful steps.

Critical Thinking Question

- What are some examples of inappropriate or rude behaviors you might experience on the part of customers?

List as many as you can think of.

Step 1: Acknowledge and Empathize with the Customer's Feelings

Some people calm down once their situation and feelings about the situation have been acknowledged. Start off your response to the customer with a sincere statement of acknowledgement and empathy. Your statements should be short and to the point. Look for things upon which you can agree with the customer.

I see that you are really frustrated by this. I would be, too, if I had paid so much money for something that only worked for the first day.

I can see that you are angry by this extra service charge. Some people feel that they pay too much as it is in service fees.

You are right. This just doesn't sound like it should. I can see why you are frustrated.

Step 2: Focus Attention on Solving the Problem

Since you can only control fixing the problem and not the customer's feelings or behavior, try to focus the customer's attention away from her feelings and toward solving the problem. Assure the customer that you are able and willing to address the issue.

I really want to help you fix this. Let's see what we can do.

I know you are really busy. I want to fix this problem for you quickly. Let me get some more details.

I am going to make this work for you. Let's see exactly what is causing the problem.

Step 3: If Needed, Set Limits

Some customers continue to be abusive, to use inappropriate behaviors, to make inappropriate statements, or to be uncooperative. In these extreme cases, you may have to set limits of behavior. Remember that setting limits should only be used if absolutely necessary and in extreme cases where a customer is being abusive or inappropriate. Be very specific on the customer's behavior that is not acceptable and on the consequences of the customer continuing that behavior. Be very calm and unemotional in your limit setting statements.

> *If you continue to swear at me, I will have to end this conversation until you have calmed down and can discuss the matter in a professional way. OK?*
>
> *If you continue to yell, I will have to come back at a different time when we can work on the problem at hand in a reasonable matter. Shall we reschedule?*
>
> *I will be happy to help you provided that you refrain from making racist statements. If you do continue, I will have to leave. It is up to you.*

Provide a choice for the customer. They can choose to continue their poor behavior or address the real problem at a later time, or they can stop the poor behavior and address the problem now. It is important not to use limits as punishment or to change the person's point of view. Your only goal should be to control the situation so that you can meet the needs of the customer.

Step 4: Do Not Respond to Intimidation

Some customers may try to intimidate you to get what they want. For example, they may threaten to go over your head, to talk to your supervisor or the president of the company, to take the issue to a newspaper or news station (the press), or to cause you or the company harm. Most people are simply trying to manipulate the situation to get what they want. In some cases, however, they may be so angry that they really do intend to do what they threaten. In any case, do not respond to intimidation. Let's look at each of these situations independently.

Customers Who Threaten to Go Over Your Head. Simply assist the customer in doing so in a matter-of-fact manner. Ask the customer to wait while you find your supervisor or manager. If they are not available, ask if you can set up an appointment for a later time. In doing so, you show the customer that you will not be intimidated. The customer will either allow you to do this or back down and allow you to help to solve the problem. Most companies have a procedure you should follow to make these kinds of appointments. Be sure to know and follow your company's policies.

Customers Who Threaten to Go to the President of the Company. Most presidents (and other high positions in the company) typically are not the appropriate individuals to solve the customer's problem at first. When a customer threatens to go to the president of the company, state to the customer that there are more efficient avenues for getting the problem fixed and offer to proceed with one of these options. If the customer still wants to talk with the president, give them the appropriate information. Follow the policies of your company as before. In most cases, the customer will begin to see that calling the president will probably not fix the problem and that the president will probably just get someone in the correct department to fix the issue. The point here is to show the customer that you will not be intimidated by threats and that you sincerely want to help fix the problem.

Customers Who Threaten to Go to the Newspapers or News Stations. With the many consumer advocates in the media, it is common for customers to threaten to go to the press to intimidate you into giving them what they want. In these instances, use the strategies already discussed to try to fix the problem. If the customer still states she is going to the press, simply state that it is her right to do so, but that you have other options that will probably solve the problem faster and easier. Ask the customer if she would like to hear about the options. If the customer insists on going to the press, do not be defensive or show that you have been intimidated. State again that she has the right to go to the press if she chooses.

Unreasonably Demanding Customers. Unreasonably demanding customers pose a very common situation. This situation is almost always the result of poor definition. Customers are usually very bright and are often sophisticated business people. They can often approach a situation as a win-lose business deal. The obvious way to avoid this is to use comprehensive and sophisticated business documentation to define the scope of the work. If the customer is reluctant to proceed under these constraints, it is often best not to enter into a business relationship. Unfortunately, this situation often develops after the project has begun. The best way to deal with unreasonable demands is to provide clear documentation supporting your position. Many of these customers will back down when they realize being unreasonably demanding will not serve their purposes. In many cases, they simply had to try. Some customers are, unfortunately, just the business equivalent of a bully who will back down when confronted. One company has even created a precaution they term "Exhibit H." This document delineates all of the things that are not going to be done at this time. It lists everything that has been discussed or offered and rejected. They have the customer initial that page. The use of "Exhibit H" is very effective in limiting the unreasonable demands of customers. Obviously, these strategies may be at a higher level than your position; however, it is important to be aware of the situations. In these cases, it is best to listen to the customer and refer the situation to a higher authority.

Troubleshooting and Problem Solving

One of the most important criteria for being able to provide excellent service, after a service-oriented attitude is already in place, is excellent problem-solving and troubleshooting skills. **Problem solving** is a general term used to describe looking at a situation, determining the

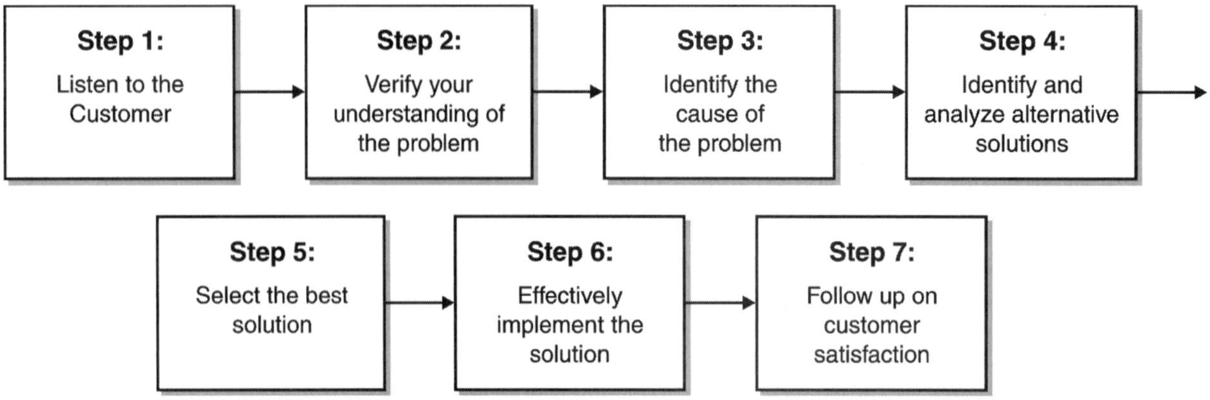

FIGURE 4–7 The Problem Solving Process

cause of the problem, developing potential solutions, and implementing the solution. The term **troubleshooting** is often used for the problem-solving activities in the technical aspects of your job, such as determining how a piece of equipment is malfunctioning. In order to provide excellent service, it is essential to be able to solve customer problems efficiently, competently, and with a great attitude.

The Problem Solving Process

Let's first begin with a review of the basic process of problem solving. The **problem solving process** is a step-by-step system for ensuring that problems are solved in a logical, thorough, and efficient manner. This process can then be adapted to almost any customer problem situation. Figure 4–7 illustrates a general problem-solving process. Depending on the situation, more or less time may be needed for the different steps.

Step 1: Listen to the Customer

Listen carefully to the customer so that you can fully understand the customer's concerns, questions, needs, and assumptions about the problem. As discussed earlier, the simple act of listening goes a long way in alleviating tempers, frustration, anger, and anxiety. Practice excellent listening skills from the very beginning. Let's review the basic elements of effective listening. These elements are explained in detail in Chapter 3 and reviewed in Figure 4–8.

Step 2: Verify Your Understanding of the Problem

Verify that you understand the problem by organizing the customer's thoughts and then restating the details back to the customer. Organizing the facts can include ordering steps, listing issues, or drawing a diagram, depending upon the type of problem.

You may also need to ask clarifying questions to the customer or someone else, provide information to help the customer understand

Skills for Effective Listening
Use appropriate body language.
Use verbal and nonverbal cues.
Avoid distractions.
Ask questions.
Don't interrupt.
Be organized in your listening.
Keep an open mind.
Pay attention to the speaker's nonverbal cues.
Analyze and minimize your own barriers to listening.

FIGURE 4–8 Skills for Effective Listening

the situation better, or carry out some initial investigating. For example, you may need to look at the system and allow the customer to show you exactly what is wrong. You may also need to make a phone call or conduct research to gain more information.

Step 3: Identify the Cause of the Problem

Once you have gathered all the initial information and conducted your preliminary investigation, you need to determine the exact cause of the problem. The cause may be a human error such as a mistake in ordering, or a technical error such as faulty equipment. You must correctly identify the cause before attempting to solve the problem to avoid wasting time in the long run. Carefully assess your assumptions. Try not to jump to conclusions prematurely. Maintain an open mind when looking for the cause of the problem. Consider all of the associated areas logically and practice critical thinking. Also use all of the resources you have available: manuals, troubleshooting guides, contractors, sales staff, supervisors, coworkers, manufacturers, and so forth.

Step 4: Identify and Analyze Alternative Solutions

There may be only one solution or there may be several alternative solutions, each with advantages and disadvantages for the customer or the company. Before making a final decision, identify all of the potential solutions and carefully analyze the effectiveness of each. What will be best for the customer? What is most cost effective for the business? What is the fastest solution? What are the needs of each player (customer, business, contractor, etc.)? What solution will fix the problem over the long term and provide the most customer satisfaction? In some cases, there may even be a short-term solution followed by a different permanent solution.

It may be that you have sole responsibility in selecting the best option. In many cases, however, you may want to present the best options with the advantages and disadvantages of each to the customer, your supervisor, the affected contractors, and others. Remember that honest and clear communication is almost always the best strategy. And, your customers will no doubt appreciate the thought you put into your problem solving.

Step 5: Select the Best Solution

After your careful selection of solution alternatives and your critical analysis of each, select the best option in light of the prioritized needs of the customer and the business. In some cases, someone else may make the selection for you. When you decide how you are going to solve the problem, you then need to explain the solution clearly to the customer. Outline exactly what you are going to do, how long it will take, how much it will cost, how it may inconvenience the customer, and how the customer will be satisfied in the end. Gain the customer's support of your solution before moving on. This clear communication will help the customer feel that you are diligently seeking to provide excellent service.

Step 6: Effectively Implement the Solution

Next, implement the selected solution quickly, efficiently, competently, and with an excellent attitude. Providing excellent service to customers means that you address customer problems quickly. It is common, in many industries, to place the fixing of a customer's problem at a lower priority than making the initial sale. In reality, it is more important to keep existing customers satisfied in order to keep the initial sales coming. Remember that unhappy customers spread their unhappiness quickly and to many people, ultimately hurting business coming in the front door. Satisfied customers will also tell others about how they were served in a problem situation. Since so many companies seem to struggle in this problem-solving area, you can set yourself and your business apart from the rest. Customers expect problems; how you fix their problems can make all the difference.

Step 7: Follow Up on Customer Satisfaction

The last important step in problem solving is to follow up with the customer to ensure that the problem has been fixed and that the customer is satisfied. Follow-up is critical to customer satisfaction and can provide you with additional opportunities to provide excellent service and develop customer loyalty. Following up with the customer can also provide you with important information to use with future customers. Review with the customer what has been done, document your customer's satisfaction (or dissatisfaction), and address any remaining issues. Consider following up immediately as well as after a longer period of time to ensure that any fixes remain fixed and that the customer remains satisfied. Follow-up also provides an opportunity for educating the customer and even making additional sales. In some cases, consider writing the customer a personal note thanking them for their business and for working with you to make the problem situation right.

Conclusions for Difficult Customers and Customer Situations

As with many things, there are proven techniques to approach and resolve almost any situation. Define the problem, consider the solutions, and select one. The best way to resolve any customer problem is to listen carefully, be empathetic, take responsibility, and create a list of steps leading to the solution that both parties agree to with the understanding that when the steps have been completed, the situation has been resolved. The obvious next step is to rigorously take action on the steps and ensure that they are quickly and effectively accomplished. The secret is in creating a finite set of tasks and gaining agreement that the problem will be solved when the tasks have been completed. It is also important to act quickly to minimize your costs.

Summary

Chapter 4 discussed important service strategies for dealing with difficult customers and difficult customer situations. A basic problem solving process was explored followed by ideas for adapting the process to various common customer-problem situations. We explored steps for dealing with schedule- and delay-related problems, customer complaints, employee mistakes, negative customer attitudes about the company, not knowing the answer to a question or problem, and inappropriate actions by customers.

In this chapter we learned:

- Problem solving is a general term used to describe looking at a situation, determining the cause of the problem, developing potential solutions, and implementing the solution.
- By following the problem solving process you can solve problems in a logical, thorough, and efficient manner.
- How you handle difficult customers is very important in meeting their needs, making them feel good about doing business with you, developing and maintaining customer loyalty, and ensuring future business.
- The first strategy in dealing with schedule- and delay-related problems with customers is prevention.
- Most customers walk away from you (usually to your competition) without saying a word and without giving you the opportunity to correct the problem; therefore, dealing with customer complaints is a great opportunity to provide excellent service to customers.
- Mistakes happen; dealing directly and effectively with mistakes is critical for excellent service.
- A customer with a bad attitude about your company based on past experiences provides you with an opportunity to provide excellent service and change the negative attitude into a positive one.
- You cannot always have all of the answers to questions or all of the information. Providing excellent service is finding the information so that the customer's needs are met.
- Occasionally, customers display inappropriate behavior and even make threats. It is important to deal with these issues tactfully and professionally.
- One of the most important criteria for being able to provide excellent service is excellent problem-solving and troubleshooting skills.

Key Terms

Problem solving Problem solving is a general term used to describe looking at a situation, determining the cause of the problem, developing potential solutions, and implementing the solution.

Problem solving process The problem solving process is a step-by-step system for ensuring that problems are solved in a logical, thorough, and efficient manner.

Troubleshooting Troubleshooting is often used for the problem solving activities in the technical aspects of a job, such as determining how a piece of equipment is malfunctioning.

Review Questions

1. What are the main steps in the problem solving process?
2. Why is it important to handle difficult customers and difficult customer situations with excellence? List several reasons.
3. What specific steps should you take to deal with a schedule- or delay-related problem with a customer?
4. What specific steps should you take when you or a coworker has made a mistake with a customer?
5. What steps should you take when you encounter a customer who has a negative attitude about your company based on past experiences or the reports of others?
6. What specific steps should you take when you don't have the answer to a customer's question?
7. What steps should you take when a customer threatens to go over your head to your supervisor or other higher position in the company?
8. What steps should you take when a customer threatens to go to the president of the company (or some other equally inaccessible person)?
9. What steps should you take when a customer threatens to go to the newspapers or news stations?
10. What steps should you take when a customer threatens to harm you or the company?

Training the Customer

After studying this chapter, you should be able to:

OBJECTIVES

- Explain strategies for teaching customers simple skills.
- Describe key principles that can be used in any training situation.
- Identify ways to verify customer understanding.

OUTLINE

Introduction

One of the important services you might be asked to provide for your customers is that of training them to use the equipment and technologies. Training is sometimes challenging. Effective training requires knowledge of your customer as well as knowledge of some basic training principles. In addition, there are many strategies that can be used to make learning easier and more effective. This chapter discusses important training principles and simple training strategies.

Types of Customers

Once your customers have purchased and installed a technology-related system into their home or workplace, they must understand how to effectively use the system to accomplish the tasks for which it was designed. As a technician, the customer will probably see you as the expert on the product. Therefore, you must not only know how to install, troubleshoot, repair, and maintain the product, you must also know how to train your customers in proper use and maintenance so that they can use the product efficiently and effectively. This training task may not be a significant part of your job; however, you will no doubt find yourself in training situations on a regular basis. Learning to train your customers with skill is one element in providing excellent service to the customer.

Teaching your customers to use their new equipment or technology can present many challenges. Let's explore a few of the possible types of learners you may face and review some strategies that may make your training situations easier.

Technically Challenged Customers

Many customers are technology challenged, meaning that they are not comfortable with using high-level technology products. These individuals may be afraid of using the product, afraid of damaging the system with a mistake, or just lack a basic aptitude for learning this kind of information. For these customers, programming the VCR or answering machine that has been in their home for decades is still daunting. They may not use a computer often and do not have the basic understanding of how any of the technologies work or even what the products are designed to do. They may be fearful of the seemingly endless pages of user manuals, features, remote controls, and so forth. They may lack confidence in turning on the product, not to mention programming it or using it to perform a function. In short, you may find that many customers are resistant to learning new skills related to technology. Even if your customer is technically challenged, it is essential that you are not condescending or do not make them feel stupid.

Training strategies for these individuals:

- Make them comfortable with their product
- Teach them how the technology works in simple terms
- Increase their self-confidence with the technology
- Teach them the basic functional features so that they can use the product
- Teach them what to do in case of a problem or question
- Teach them how to find answers for themselves in easily accessible resources such as the user's manual or the Internet.

Critical Thinking Questions

- What are some signs that a customer is technically challenged?
- How might you detect this if they do not come right out and tell you?
- What might be some of their specific fears?

List as many as you can.

Technology Competent Customers

Many of your customers will be very comfortable and competent with technology products, but they may not know how to use your specific products or system. These individuals may learn quickly, but may be difficult to teach. They may think they already know how to use the product, may avoid user manuals or instructions, or may feel as though they do not need to waste time in training. In reality, they may not know as much as they think they do and can run the risk of damaging the product or in missing out on some of the functionality.

It is important to deliver information without forcing the customers to acknowledge that they do not already know something or to claim to know something when they actually do not. Pride is your adversary. Be careful to avoid situations where a customer's pride can be damaged. And use pride to your advantage. Look for opportunities to allow the customer to shine. For example, if you were explaining a surround processor you would not ask, *"Do you understand how the technology works?"*. You would not get a favorable result, either the customer will claim to know it and you will not have the opportunity to explain or you will have set up a situation where the customer is embarrassed. The effective way to deliver information is to say, *"You probably already know how a surround processor works, it works like this . . ."* and then proceed to say what you want to say. Simply put, give the customer credit for probably knowing it and then tell them what you want them to know.

Training strategies for these individuals:

- Acknowledge their existing competence and avoid being condescending
- Ensure that they can use the basic features and make them aware of the more advanced features
- Teach them advanced skills, shortcuts, maintenance tasks, and troubleshooting strategies so that they can be as self-sufficient as possible.

Critical Thinking Questions

- How might you tell whether or not someone is technically skilled if they do not directly tell you so?
- What problems might this type of customer cause that would be different from those of a technically challenged customer?

Training Needs of the Family

You may be asked to train an entire family to use a product, including children of all ages. Children tend to learn differently than adults. Teens lie somewhere in between children and adults depending upon their age and maturity.

Training strategies for these situations:

- Determine the user goals for each family member and adjust your training focus accordingly
- Discuss the user goals of the children with parents

FIGURE 5–1 Training needs will be different for each member of the family. Understanding and implementing various training strategies will help you become a successful trainer.

- Ensure that you use an appropriate training strategy for each user
- Teach appropriate skills for parents to fully monitor and control the use of the product by children, if necessary.

Critical Thinking Question

- In what situations might you need to train kids in the use of equipment?

Think about your own job situation or a job you would like to have in the future. Be as specific as you can.

As you can see, you need a wide variety of training strategies in your toolbox to meet all the potential training needs that come up in your job. Let's review in more detail some of the challenges that different types of customers might present. The more you understand about your training audience, the better you will be able to meet their learning needs and the better service you can provide to them.

Key Principles for Training Situations

Becoming an effective trainer involves developing several skills based on established training principles that have been proven in many different types of settings. By following key training principles, you will be able to successfully train your customers in simple skills they will need to use their equipment.

Fear Is Paralyzing

This powerful emotion can significantly hinder learning. The more you reduce the fear of a product for the customer, the easier it will be for your customer to learn. Fear is often associated with resistance to learning something new. Many adults, especially older adults, think that they are too old to learn something new, especially when technology is involved. Although this is a generalization, it is important to be aware of the possibility.

Training strategies when fear is an issue:

- Reduce or eliminate the fear
- Encourage the customer to explore and become comfortable with the product
- Increase the customer's level of confidence with the equipment
- Explain the equipment so that the customer understands how it works. (Fear is often caused by a lack of knowledge and understanding.)
- Allow the customer to practice while you are there to help if they get stuck.

Different People Have Different Learning Styles

Learning styles describe the way in which an individual learns best in most circumstances. Some people learn mainly by hearing (**auditory learners**). Some people learn mainly by seeing diagrams and images (**visual learners**). Some people learn mainly be using or doing (**kinesthetic learners**). Most of us learn best by some combination of these three learning styles. The more you can utilize elements of each different learning style, the more effectively you will reach your customer. Also, though most people utilize some elements of each learning style to some degree, we typically emphasize our main learning style when stressed. Learning new skills can be very stressful. Figure 5–2 illustrates the basic learning styles and how they overlap in different individuals.

You may not know exactly which is the best learning style for each individual customer; however, individuals often present clues to help you make a good guess. For example, people who talk to themselves or verbalize instructions as they are learning are probably auditory learners. People who draw out instructions or who draw pictures of control panels to help remember instructions are probably visual learners. People who go immediately to the control panel or piece of equipment are often kinesthetic learners. Try to pick up on these cues and utilize appropriate teaching strategies. For example, if you deduce that your customer is a visual learner, draw out your instructions on a piece of paper or use the diagrams in the user manuals. If you deduce that

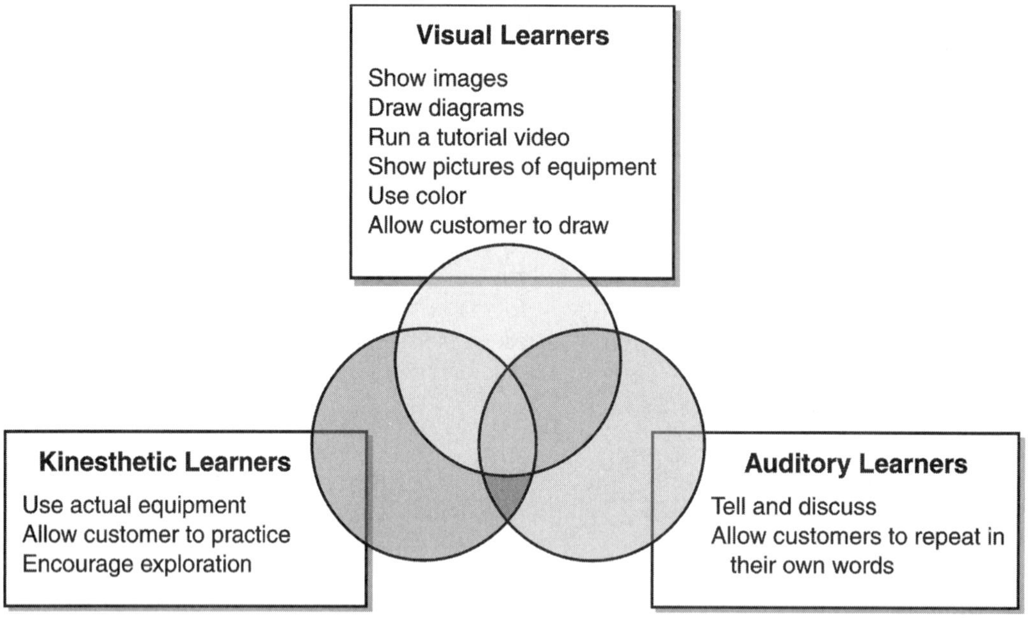

FIGURE 5–2 Learning Styles

your customer is a kinesthetic learner, use the actual equipment to explain the features. If you see that the customer is an auditory learner, discuss how the equipment works.

Important Note: Most people tend to teach using their own learning style. For example, trainers who are visual learners tend to teach using pictures and diagrams. Be aware that your customers may not learn in the same way that you teach. It is important to understand your own learning style and to work on developing other strategies to assist customers who learn differently. Your goal is to adapt your teaching to meet your customers' needs.

Training strategies to effectively match the learning style of the customer:

- Try to determine which learning style the customer relies on most
- Determine what your own learning style is and learn how to communicate using other styles effectively
- Ensure that you include strategies from all learning styles in your training
- Think about ways to teach common skills in all different styles before you meet your customers.

Teaching Adults Is Different than Teaching Children

Adults and children are very different in how they think, especially in how they approach learning something new. There are other important differences between children and adults that should be considered in a training situation. Understanding these differences helps to not only apply appropriate strategies to each learner, but also to avoid certain strategies. Treating a child like an adult in a learning situation tends to result in poor learning. Treating an adult like a child tends to cause offense.

On the other hand, children often approach learning new things with enthusiasm and confidence. They do not yet have that many unsuccessful learning experiences and are often familiar with computers, computer games, and other technologies. Adults may not have grown up with the breadth of technology as our children do today. Adults may also be more aware of the risks of failing. For example, adults may assume that if they break a system, they will have to pay a service fee to repair the damage. These are broad generalizations and should be addressed carefully. Many adults are very comfortable with technology, and some children are intimidated by it. The best strategy is to approach both children and adults with an open mind, but be aware of all the possibilities and how to deal with them. You might even directly ask your customers how familiar they are with their system. You never want to offend your customers, but you want to set them at ease with the technology from the beginning.

Adults Tend to Apply Information Learned in One Situation to a New Situation

Children are typically sequential and categorical. In other words, children tend to learn new concepts or procedures more easily if presented step-by-step. And, they keep concepts in categories without easily applying an idea from one category to a different situation. On the other hand, adults more easily apply steps learned in one situation to achieve goals or follow procedures in a new situation. And, although many adults learn information best if it is taught in steps, they are more adept at applying information out of sequence and to new situations.

Adults Prefer to Be More Self-Directed in Their Learning than Children

This means that adults respond less readily to the teacher-student situation. Children are used to being in this situation daily in school where the teacher is the authority and expert. Adults often prefer to be acknowledged for existing knowledge and experiences, would rather control the learning situation, and resist environments where the trainer assumes any kind of authority over the learning situation.

Adults Find It More Difficult to Memorize Information than Children

In general, adults dislike memorization. They prefer to understand why they must do something to achieve results than to learn by rote memorization of a procedure. So, for an adult, it is more important to explain the "whys". For children, it is better to just explain what needs to be done.

Adults Have Many Experiences from which They Can Draw

A good strategy for teaching adults is to compare what they are doing with something they already know. For example, if an adult already has a security system in the home and you are explaining the operation of a new integrated system, you can easily compare and contrast the operation of the new system with the old system. Depending upon the age and experiences of the child, this teaching strategy might not work as well. For children, you may need to just explain the operation or appropriate steps without comparing it to anything else.

Adults Are Problem Oriented

Your adult customer has purchased a system to solve specific problems or achieve specific goals. Therefore, the training needs be directed to solve the problems or meet the goals. Children, on the other hand, typically are not as focused on these problems. They need more guidance in what the system can actually do for them. This ties back to the need for more directed learning for children. The goals of children relate more to having fun and being entertained.

Adults Care about Value

They need to see that the products they are purchasing are of high value. You can show value by explaining to the adult how the product or service can be used to make life easier or solve problems. The more sophisticated the user, the more they tend to appreciate value-added elements such as shortcuts, services, advanced features, possible upgrades, and so forth. Remember that children care little about value; they just want to have fun.

Children View Having Fun as More Important than Learning

Adults will judge their training experience by how well you meet their training needs, how well the information is presented, and how they perceive the value of the training compared to the cost of the product and service. Obviously, adults will probably expect a higher quality training experience for a very expensive system. Children, on the other hand, will judge the training experience by how well they relate to the trainer and how much fun they have. Remember that, although your customer is the adult, it is important to provide excellent service to the children and to meet their goals as well.

Children Are More Trusting of the Trainer

Adults can sometimes be skeptical about the accuracy of instructions, about the actual functionality of the system, and about the trainer's knowledge of the system. No doubt this is because of many previous disappointments. However, children will typically believe what the trainer says since the trainer is the adult, is the perceived expert, and is supposed to know what they are doing. They have not yet experienced as many disappointments in sales people, products, and services. Adults often need to test the product for themselves by actually following the directions and seeing what happens.

Study/Career Tips

Before any training situation, think through what we know about adult learning and how it applies specifically to the customers you will be training.

- Adults sometimes think that they are too old to learn new things.
- Adults tend to apply information learned in one situation to another situation.
- Adults prefer to be more self-directed in their learning than children.
- Adults find it more difficult to memorize information than children.
- Adults have many experiences from which they can draw.
- Adults are problem oriented.
- Adults care about value.
- Children view having fun as more important than learning.
- Children are more trusting of the trainer.

Teaching Procedural Learners vs Teaching Navigational Learners

Massie (1997) presents a useful distinction between different types of learners. He explains that in computer learning, people are procedural or navigational. **Procedural learners** memorize a series of steps to achieve a specific function. They apply these steps exactly any time they want to perform the function. **Navigational learners** understand the concept and reason behind each procedure and can reapply a set of procedures, perhaps in a different way, to perform a different function as the situation requires. Once procedural learners learn a procedure, they tend to require little assistance in completing the task as long as everything works as expected. However, they tend to resist experimenting for new functions and are often lost if something goes wrong. On the other hand, conceptual learners tend to spend a lot of time experimenting to find out all the possible things a system can do. And, they tend to get into more trouble while experimenting.

To train a procedural learner, you must emphasize the exact steps. You can even write or draw out steps for each function to be performed often. To train a conceptual learner, teach the basic functionality and then focus on the user manual, Web site, or other resource that will help them in their experimentation.

Teaching Strategies for Training

There are numerous strategies for effective training. Let's explore two simple strategies that may help your training situations run more smoothly: the Preview-Teach-Review Model and the Cognitive Apprenticeship Model.

The Preview-Teach-Review Model

A very basic teaching strategy that works for most teaching situations and populations is the **Preview-Teach-Review Model.** It encourages a trainer to follow three simple steps: (1) preview the information, (2) teach the information, and (3) review the information. In short, tell the learners what you are going to do, do it, and then tell them what you did. This simple strategy can be combined with other strategies to achieve good learning results. It allows the learner to define the big picture before moving into the details and then to review the big picture after learning the details.

Preview the Information. Preview means to provide a brief introduction explaining to the learner the goals of the training session, what you will be doing, why it is important to learn, and how the training will be conducted. By providing the learner with a view of the big picture of the training goals, the learner is better able to put the details in context.

> *The purpose for this feature on your security system is to set the system so that you can move freely throughout the house, but still be signaled when a door or window is opened to the outside. This process is done in four simple steps that we will go through together.*

Teach the Information. Next, explain the concept thoroughly. Explain how the technology works and how it is connected together to other systems in simple terms, if applicable. Explain the features and parts of the equipment. Go through the steps of the process. Give the learner tips for remembering steps. Watch the customer's reaction to get an idea for their understanding. Note common problem areas and how to get around these problems.

> *Let's look at each of these four steps. The sequence is important. First, hit the control button. Next, . . .*

Review the Information. Finally, review the information again with the customer. Go through a brief review of the steps and equipment. Ask for any questions. Provide answers as necessary. Allow the customer to complete the tasks themselves while you provide feedback and guidance.

> *So, what we did was to set the security system so that you can move freely about your house, but will be signaled if a door or window is opened to the outside. This feature is enabled in four simple steps . . .*

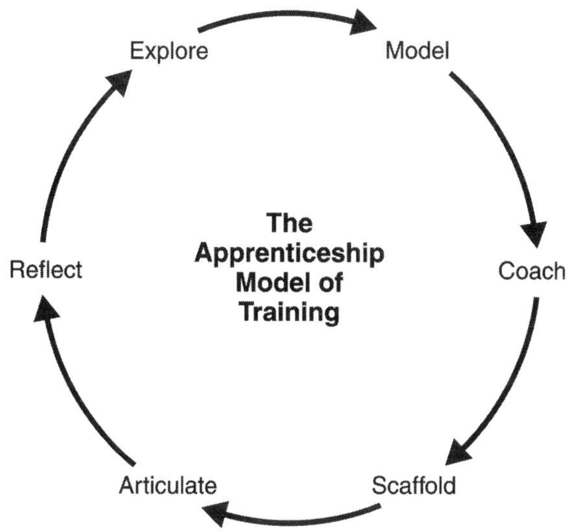

FIGURE 5–3 The Cognitive Apprenticeship Model

The Cognitive Apprenticeship Model

Teaching hands-on skills can be accomplished effectively using an apprenticeship-type model of training. Apprenticeship has been used for centuries to teach people how to do skilled tasks. Even today, many different industries train workers by using apprenticeship experiences. The apprenticeship-type model of training illustrated in the **Cognitive Apprenticeship Model** (Collins et al. 1991) is an effective step-by-step approach to teaching hands-on skills. Figure 5–3 shows the Cognitive Apprenticeship Model.

Model. Start with an explanation of the equipment, controls, or tools. Next, demonstrate how the procedure or skill is to be correctly performed. Explain the final goal, explain each step as you complete it, and demonstrate that the steps actually achieve the goal. Be encouraging, positive that the customer can learn the procedure, and slow enough to allow the customer to focus on the steps. Start out with an encouraging introduction to the task.

> *Here, let me show you how easy it is to make a recording of the TV program. This is the control panel for recording. This button starts the recording. This button stops the recording . . .*

Coach. Next, ask the customer to complete the task just as you have demonstrated. Give instructions at each step as needed, compliment the successful completion of each step, and maintain your positive attitude that the customer will be successful. Give sufficient hints to allow the customer to complete the task successfully. Provide reminders of commonly forgotten steps or problem areas.

> *Now, you try. First set the time to the current time.* (Allow customer to set the time.) *Excellent! Next, . . .*

Scaffold. Scaffolding means to start with simple tasks and then move to more difficult tasks as the customer gains knowledge and confidence. This means that it is best to start with the simple procedures or simple aspects of the procedures before moving on to advanced or more difficult tasks. The goal is to develop basic skills and confidence in the customer as they progress to more difficult tasks. Choose lower risk aspects of a task so that if the customer does make a mistake, the problem can be easily remedied.

> *Let's start with something simple. We can just hit "reset" to move it back to its original status. Then we will move on to the more complicated features.*

Articulate. Next, review the details of the procedure with the customer. Consider asking the customer to recall the steps by themselves. Provide any correction or expand on their recalled information as needed. If the customer is having difficulty, ask them to verbalize what they should do at each step. This strategy will help them organize the task and determine where they are stuck.

> *OK, now let's review. If you want to record a TV show, where do you start?*

Reflect. Allow the customers to reflect on what they just learned. Some individuals may take a little longer to go through the process and decide if they are comfortable with it. Some customers may want to take notes and then review them to ensure they have written everything down correctly.

Explore. Finally, allow the customers to explore the system on their own, either while you are with them or after you have gone for the day. People will more readily explore their systems if they are confident that you will be available to help them as needed or to fix anything they mess up. Show them how to use resources, such as the user manual or a Web site with user information.

> *OK. Now I want you to experiment with the system and see what it will do. Here is my cell phone number. If you run into trouble or have a question, don't hesitate to call. Here is the correct page in the user manual. Let me show you how it is organized.*

Verify the Learner's Understanding

Before leaving the training situation, it is important that you verify that the customers understand everything they need to understand to get started with the system and to ensure immediate success. Some people are shy and will not ask questions. Others may be embarrassed that you have spent time training them yet they still do not understand. Still others may be so lost that they do not know where to begin to ask you questions.

There are several techniques you can use to ensure that your customers understand enough to be immediately successful in at least the basic functionality of the system or equipment.

Ask Your Customers Directly If They Understand

Many people will answer you truthfully. However, some people do not want to look stupid and some may not even know if they really understand until they have to perform a task without your help. Be supportive. Ask if they have any questions. Ask if them if you have been clear in your explanation. By taking responsibility for their learning, you may make it more comfortable for them to tell you what they do not understand.

Look for Nonverbal Cues of Misunderstanding

Watch your customers carefully during the training and when you ask questions. Do they wrinkle their brow? Do they avoid eye contact with you? Are they deep in thought? Are they reviewing any notes they have taken? Are they scratching their head or touching their face? Do they appear impatient or frustrated? Try to determine through behaviors and expressions what you need to do next.

Leave Behind a Cheat Sheet

If the information is new, your customers may not be able to remember everything you say in your training. Leave a cheat sheet of simple instructions. You can write these down as you go through them, or even have them already printed up before you arrive.

Follow Up Shortly After the Training

As a courtesy, follow up with your customer shortly after the training. Some people need time to digest the information, try out the features or equipment by themselves, and determine what they do not understand. Many people will be hesitant to call and may appreciate a short call to see how they are progressing. Consider making the call a day or two after your training. Before you leave the training session, tell the customer that you would like to check in with them and ask them what a good time would be. Encourage them to write down any questions or problems they have. Also, leave your contact information so that they can follow up with you as well.

General Tips for Excellent Training

Regardless of the strategy you choose for your training, the following tips can be used to ensure that you have explained the information clearly and have provided the optimal learning environment for your customers.

Plan Ahead

Review the training material or the equipment prior to meeting with your customer. A quick review will ensure that you are clear in your explanation, that you train your customer with confidence, and that you give a professional impression. Be sure that you have any cheat sheets you may be leaving with the customer.

Speak Slowly and Clearly

Speak slowly and clearly in your explanation. Many people speed up when they are nervous or when they are very confident. Think about your delivery. The information may be new to your customer and terms may be unfamiliar. Speaking slowly gives them an opportunity to digest what you are saying as you go along.

Be Patient

Patience is critical in any training situation. Remember that your customer may not pick up on topics or skills that you think are easy. They may have different aptitudes or be fearful of the technology. Both words and nonverbal cues reflect impatience. Watch carefully for responses, facial expressions, tone of voice, body language, and so forth.

Smile and Maintain Enthusiasm for the Task

A sincere smile is an excellent way to encourage customers in their learning. Also, your enthusiasm will be contagious. If you are bored with the information or product, your customer will pick up on this attitude.

Be Logical

Think carefully how you will present the information. For many tasks, it is best to use sequential steps, numbered items, or a chronological flow of events. Making your training logical will assist the customer in both learning and retaining the information.

Teach in Small Chunks

George Miller (Miller, 1956) conducted studies on humans to determine how much information they could learn at one time. He found that people typically store 7 ± 2 different items into short-term memory. In other words, if asked to memorize a set of items, most people can only memorize in the range of five to nine items. To increase the number of unrelated items that can be put into short-term memory, the items must be related in some way, such as with a memory or mnemonic device like a formula or rhyme. The result of Miller's research is the concept of chunking information into small, related groups. **Chunking** means to organize information into small groups and to make the items in the small groups relate to each other in some logical way.

Allow Time for the Learner to Think

Every person thinks at a different speed. Some people must look at the entire picture and run through the process in their minds. Others pick up on processes quickly, but miss out on many of the details. Allow your customers to move through your training at their own speed. If they need a few seconds to think about what you are saying or to answer a question, give them the time. Remember that for every question you ask, the customer must understand the question, think about the process or information, formulate a response, and then verbalize the response. This may take several seconds. Let your customer think in silence. Watch body language and facial expressions to determine when to intervene or help with the response.

Explain Shortcuts

Some customers will only want simple steps in a process. For example, if you are training them to use a piece of equipment, some customers will want to follow the same steps each time they perform the function. Others, however, will want to learn shortcuts or time-saving tricks. Consider asking your customers directly if they want to learn a shortcut. You will soon find which learners do and which ones do not want to be confused with different information.

Provide Hints and Memory Tricks

When possible, provide to your learners hints on how to remember steps or specific functions or memory tricks such as using the first letter of each step in a series to form an easily remembered word. Other simple tricks include reminding learners to always start with a specific step such as *always start with the red button* or *always go back to the main menu.*

Practice as Necessary

If you are new to or uncomfortable with training others, practice before actually meeting with the customer. Try out your teaching on your friends or family or run through your presentation with a coworker. Get as much feedback as you can on your delivery and content. The more practiced you become, the better you will be able to provide excellent training to your customers.

Demonstrate Using Actual Equipment

Most people are visual learners or kinesthetic learners or some combination of the two. When possible, use the actual equipment to conduct the training. Rather than going through the user manual to read the steps of a process, take the customer to the system itself and demonstrate on the real equipment. You might want to point out the steps in the user manual so they can refer back to the instructions, if needed, but a real demonstration is typically the most effective strategy.

Use Visuals

Try to appeal to people with visual learning styles by using a diagram, the actual equipment, or even a handmade drawing as you go through instructions or concepts. When possible, use color in a logical way.

Summary

In this chapter we discussed very basic skills needed for effective training of customers in use of equipment or related concepts. Even if training is not the main responsibility of your job, you may find yourself in a training situation with a customer. You may have to teach customers to use new equipment or teach them how to troubleshoot or solve a recurring problem on their own. Teaching even simple procedures should be done in a logical, professional, and skillful way to provide your customer with excellent service.

In this chapter we learned:

- You may be faced with many different categories of learners: those who are technically challenged, those who are technically competent, and both adults and children. Each different learner population requires different strategies of training.
- There are many key principles for effective training that any person in a training situation should understand to ensure that the training is optimized.
- Fear is paralyzing for some people. A customer's fear may hinder their ability to learn how to use equipment.
- Different people have different learning styles. Individuals may prefer to learn using visual elements (visual learners), sound or words (auditory learners), or by doing things themselves (kinesthetic learners.)
- Teaching adults is different than teaching children. It is important for trainers to understand the differences and use appropriate strategies for each type of learner.
- Teaching procedural learners is different than teaching navigational learners.
- Adults sometimes think that they are too old to learn new things.
- Adults tend to apply information learned in one situation to another situation.
- Adults prefer to be more self-directed in their learning than children.
- Adults find it more difficult to memorize information than children.
- Adults have many experiences from which they can draw.
- Adults are problem oriented.
- Adults care about value.
- Children view having fun as more important than learning.
- Children are more trusting of the trainer.
- The Preview-Teach-Review Model encourages a trainer to follow three simple steps: (1) preview the information, (2) teach the information, and (3) review the information.
- The Cognitive Apprenticeship Model is an effective step-by-step approach to teaching hands-on skills that includes the following steps: model, coach, scaffold, articulate, reflect, and explore.
- Before leaving the training situation, it is important that you verify that the customer understands the main concepts and procedures of the training.

Key Terms

Auditory learners Auditory learners are those who learn mainly by hearing.

Chunking Chunking means to organize information into small groups and to make the items in the small groups relate to each other in some logical way.

Cognitive Apprenticeship Model The Cognitive Apprenticeship Model is an effective step-by-step approach to teaching hands-on skills.

Kinesthetic learners Kinesthetic learners are those who learn mainly be using or doing.

Learning styles A learning style describes the way in which an individual learns best in most circumstances.

Navigational learners Navigational learners understand the concept and reason behind each procedure and can reapply a set of procedures, perhaps in a different way, to perform a different function as the situation requires.

Preview-Teach-Review Model The Preview-Teach-Review model is very basic teaching strategy that encourages a trainer to follow three simple steps: (1) preview the information, (2) teach the information, and (3) review the information.

Procedural learners Procedural learners memorize a series of steps to achieve a specific function and apply these steps exactly any time they want to perform the function.

Visual learners Visual learners are those who learn mainly by seeing diagrams and images.

Review Questions

1. What are four strategies for training technically challenged individuals?
2. What are four strategies for training technically skilled individuals?
3. How does fear affect some individuals in terms of learning new information?
4. What is a learning style and how does it affect training?
5. What kinds of teaching strategies would you use to effectively train a visual learner?
6. What kinds of teaching strategies would you use to effectively train an auditory learner?
7. What kinds of teaching strategies would you use to effectively train a kinesthetic learner?
8. How is teaching adults different than teaching children? List at least ten differences.
9. What is the difference between a procedural learner and a navigational learner and how would you adjust your teaching strategy for each?
10. Explain the Preview-Teach-Review Model for teaching.
11. Explain the Cognitive Apprenticeship Model for teaching.
12. What are ways you could verify that a learner understands what you are training?
13. List ten tips for excellent training.

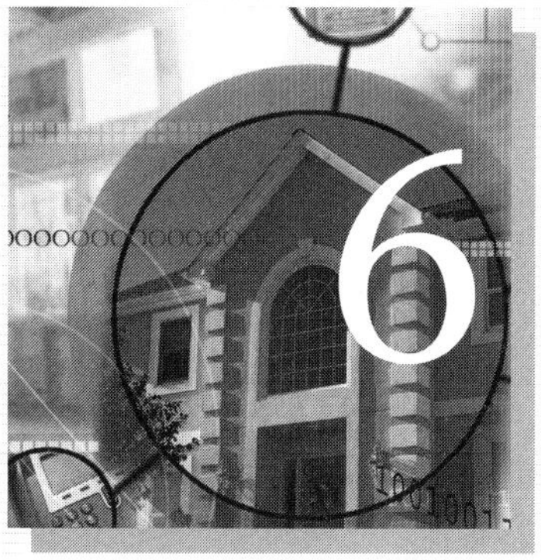

Ensuring Customer Satisfaction

OBJECTIVES

After studying this chapter, you should be able to:

- Identify effective strategies for following up with a customer.
- Explain basic sales techniques that can be used with customers.
- Identify ideas for going the extra mile for customers.
- Explain the importance of going the extra mile to ensure customer satisfaction.

OUTLINE

Introduction

Customer Follow-Up

The Next Sale

Going the Extra Mile

Introduction

To complete a customer interaction, several final steps should be taken, depending upon the situation. Some situations call for follow-up on satisfaction to ensure that a problem has been solved. Existing customers should be followed up with, in general, on a regular basis. Other situations provide opportunities for the next sale. Satisfied existing customers are often excellent candidates for the purchase of additional products and services or product upgrades. Satisfied customers also make up the foundation of a referral base, which is the only way to survive in an industry in which advertising is largely ineffective. A referral lead is often an automatic sale with no competition. In addition to being good business, it is much cheaper to maintain a strong referral base than to field a sales force in constant search of new business. In all customer interactions, it is important to maintain the attitude of going the extra mile. Although every situation is unique, going the extra mile helps to secure customer satisfaction and long lasting customer loyalty. This final chapter discusses the importance of and strategies for customer follow-up. It also explores some basic sales techniques that can be used to either sell or plant the seeds for product sales or upgrades, recognizing that not every installer is in this type of situation, or for the purpose of creating awareness of ongoing customer satisfaction and relationship. Finally, we will identify benefits of going the extra mile with customers and explore several "extra mile" ideas.

Customer Follow-Up

Completion of a job or a communication with a customer is typically not the final step. Excellent service often includes following up with the customer to ensure that everything works as it should, that the problem has been solved, and that the customer is satisfied. And, not every customer problem or question can be addressed in just one interaction. In many cases, follow-up can be successfully accomplished with a quick phone call. In other cases, a face-to-face visit is in order. Let's explore goals and strategies for customer follow-up in more depth.

Goals for Customer Follow-Up

Providing excellent customer follow-up meets several service goals:

- Follow-up tells the customer that they are valued and appreciated.
- Follow-up lets you know if the customer is happy with the equipment and installation.
- Follow-up lets you know if the customer understands how to use the equipment or if there are questions.
- Follow-up lets you know if the problem was solved or the question was answered satisfactorily.
- Follow-up allows you to assess the quality of your products and services in an informal situation.
- Follow-up allows you to assess your own work and service or that of your coworkers, subordinates, or partners.
- Follow-up provides an opportunity for additional sales (or for at least planting seeds for additional sales) of other products and services offered by your company.
- Follow-up provides an opportunity for leads or referrals of potential customers.
- A consistent habit of follow-up helps to secure your customer's loyalty.

Critical Thinking Questions

Think of time when an employee followed up with you after performing some service or fixing a problem.

- How did this follow-up communication make you feel?
- What specific goals do you think the follow-up communication met for the employee?

If you have not had this experience, think about how you might feel if a sales person who just sold you a computer or an electrician who just fixed a problem in your home called the next day to see how things were going. Write down your thoughts.

Follow-Up Strategies

There are many strategies and practices that can make following up with a customer efficient and effective.

Follow Up Enthusiastically

It is important to view following up with your customers as an opportunity to achieve the goals described previously rather than as a chore. Customers always see through insincerity and a poor attitude in a follow-up communication can easily be perceived as a nuisance rather than a sign of excellent service.

Follow Up with a Purpose

Follow up with a purpose in mind: to see if the problem was solved, to see if the customer is happy with a new purchase or installation, to see if the customer has any questions about new equipment or your recent training session, and so forth. Consider keeping a notebook of your daily tasks and communications and make notes about issues for follow-up.

Follow Up in a Timely Fashion

Effective follow-up requires taking prompt action on tasks you tell a customer you will complete, such as finding the answer to a question, submitting an order, scheduling a service call, solving a problem, etc. It is important that you think carefully about the response time you give to a customer. You must be able to actually do what you say in the time you have allowed. It is better to complete a task ahead of schedule than to be late. Also, if possible, be specific in the time you will get back to the customer. Rather than saying you will give the customer a call tomorrow, ask the customer when a follow-up call is convenient and then call at that specific time.

> *I will need a few hours to research this question, Mr. Adams. How about if I call you later this afternoon at 5:30? Or, is there a better time?*

Follow Up Using a Variety of Communication Channels

Some situations require a face-to-face interaction or visit to the home for the most effective follow-up. Many situations, however, require only a phone call or an e-mail. Give your customers the choice. Many people would rather be called only at home or only on their cell phone and do not want to be bothered at their workplace. Others prefer e-mail communications. Get to know your customer's preferences and use them when possible. Ask them directly as needed.

> *I am going to check up on how you like your system next Monday to see if you have any additional questions after using it for a week. When is the best time to give you a call? Or, if you prefer, I can e-mail you.*
>
> *Shall I phone you or e-mail you the equipment ordering information when I get it? Or, you can call me at your convenience when you get home.*

Follow Up on Customer Complaints

A complaint actually provides an opportunity to show how much you value your customer. Follow up immediately when you learn that a customer has a complaint. It is thought (though very difficult to accurately measure) that for every one complaint, there may be as many as 25 other customers with complaints who do not speak out. These customers just leave or never do business with the company again. By following up on a customer complaint, you can (1) fix that customer's problem, and (2) see what problems other customers may be having, hopefully to fix the problem for everyone. Also, it appears that a customer who has had a problem quickly and successfully resolved is more likely to return to or stay with the company for future business. Remember that a customer who has a problem will tell as many as ten people about the problem. They will also tell others if the problem was resolved quickly.

Unfortunately, not everything you do will be perfect, and equipment is sometimes defective. Everyone expects to have problems at some point. How you deal with these problems means the difference between a customer lost to the competition and a loyal customer. Use your problem-solving skills to effectively and efficiently fix the problem. Thank the customer for their patience and business. Finally, follow up again to ensure that the problem has really been fixed and that the customer is satisfied.

Follow Up at Every Opportunity

It is obviously important to follow up with your customer after you have completed a service or fixed a problem; however, it is also beneficial to follow up periodically with your customers even several weeks or months after the service. This extra thought assures the customer that you are still thinking of them and that you still care about their needs. A one-minute phone call to see how a customer is doing can make a huge impression on your customer and open the door to additional purchases and referrals.

You may also be able to solve simple problems or answer simple questions over the phone, securing customer loyalty for the long-term.

A large majority of customers who do have a problem or complaint never say anything. They just go to your competition to do their business. By following up even without a complaint, you create an opportunity to make a problem right and save the customer. The customer may also give you valuable feedback that you can use to make your service, products, or company better. Even if everything is perfect, you have made the customer feel appreciated and valued.

Follow Up on Your Promises

Obviously, if you say you are going to follow up, do so. There is nothing that ruins a relationship better than not doing what you say you are going to do. Make sure that you can fulfill your promise before making it. Take notes so that you remember what you are to do in detail.

The Next Sale

One important goal of excellent service is to ensure that the customer continues to look to your company to meet their product needs in the future. If they need more of the product, or a different product, or an upgrade of a product, they should meet these needs through your company. Although sales is probably not one of your primary responsibilities as listed on your job description, you are nonetheless in a type of sales position. First, the impression you leave on your customer influences their future purchases with your company. Second, excellent employees constantly look for any sales opportunity for the benefit of their company. Any employee who has contact with customers potentially has an opportunity to facilitate the next sale. To achieve this goal, it is essential to have a thorough understanding of the potential sales opportunity presented by each customer.

To thoroughly understand the sales opportunity, you must understand

- The customer's existing problems, needs, and interests
- The customer's potential future problems, needs, and interests
- The products and services offered by your company
- The products and services your company will soon offer, even though they are not currently available
- Use of effective selling strategies
- Your company's sales procedures to ensure that sales opportunities are facilitated effectively.

Selling does not require that you be pushy or aggressive. You should definitely focus on your primary job. However, every customer interaction may present you (your company) with an opportunity for an additional sale. It is important to be aware of these opportunities and to be equipped with the knowledge and skills to move the customer toward the purchase.

Let's look at each of these areas in more detail.

The Customer's Existing Problems, Needs, and Interests

In order to interest an existing customer in purchasing new products and services, you must know what they have already purchased and their areas of interest. The more information you have about your customers, they more accurately you can direct them to appropriate products. For example, a customer who has recently purchased a home entertainment center may be very interested in technology that downloads movies from their cable service provider. By (1) knowing your customer's current situation, (2) understanding the existing technologies, and (3) explaining the technologies to your customer, you may very well sell them additional products and services as well as add value to their recent entertainment system purchase. By explaining the available technologies, you set the stage for equipment upgrades down the road, for more equipment, for additional or upgraded services, and so forth. They may not purchase anything from you immediately; however, you have planted seeds for the future. Remember that it benefits you to assist your own company and your partner companies in selling their products. And, by helping your customers stay satisfied, you develop customer loyalty for the future.

Dale Carnegie (1888–1955) was a master at some of the basic concepts of selling. In his book *How to Win Friends and Influence People* (Carnegie, 1936), he outlines many concepts on human relationships that apply well to the employee-customer relationship, especially in sales situations. His main premises revolve around the fact that everyone is different and that you must first view things through their eyes in order to influence them to your way of thinking. To do this, you must practice empathy, understand their needs and specific situations, and communicate with them on a personal basis.

The Customer's Potential Future Problems, Needs, and Interests

In many cases, customers purchase a product for a specific purpose and do not think about the possibilities of the future. Think about the following simple illustration. A technically challenged customer purchases a laptop computer from your company simply for the reduced space requirements in the home and for the ability to move it easily from room to room. The customer has no immediate need for mobile technology. In your relationship-building communication with the customer, you find out that the family is planning a two-week vacation. You briefly bring up the concept of wireless technology and how their new laptop can easily be adapted for wireless capabilities. You talk about the availability of wireless hotspots in hotels, on-line movie and game vendors that allow inexpensive downloading of movies (compared to the expensive movie and game purchase in hotel rooms), the easy connection of the customer's digital camera to the laptop for immediate downloading and sending of photos to family, and so forth. Although the customer may not have these needs currently, you have planted seeds for the future by providing simple information and explaining the capabilities of the new laptop.

The Products and Services Offered by Your Company

Obviously, to provide useful information, you must know your company's products and services well. You must also know the products and services sold by your partnering companies. Product knowledge is critical, even if you do not deal with the products directly. Product knowledge is gained by ongoing study. Most business now have their products explained thoroughly on the company's Web site. The company no doubt has numerous marketing tools. You can also gain valuable information by reading manufacturer information and technical or user manuals. By learning as much as you can about the capabilities of your company and partnering companies, you not only equip yourself to provide important information to your customers and potentially plant the seeds for additional purchases, you also show your company that you are a committed employee.

It is important to know the following information about your products and services:

- The products your company sells and the name of the product as described in marketing pieces
- The differences between the different models
- The latest model of a given product and any improvements that have been made over previous models
- The features on each product and how these features work
- The options available on products (size, color, speed, power, resolution, sound, etc.)
- The cost of the products
- The approximate time it takes for delivery and how deliveries are made
- How products are installed or hooked up
- Maintenance needs for products
- How products work together with other products
- How products work (in simple terms that anyone can understand) in theory
- How to demonstrate the products, including all features
- Some of the downsides of the products and how these can be overcome
- Other services that may be needed (cable, phone, high-speed Internet, etc.) and their ballpark costs
- How your products differ than your competitor's products.

Study/Career Tip

Go to the Internet and find an electronic system product you are unfamiliar with.

- What kind of information do they provide about their products and services?

Make a list of all resources that you might use to improve your product knowledge. Compare your list to the lists of others.

The Products and Services Soon to be Offered by Your Company

Most companies are planning today for the technologies of the future. It is helpful to gain insight into where your company is moving next in terms of new products and services. Although you may not be able to make an immediate sale to your customers, you can plant seeds and develop in your customers an interest in future products. By being excited and informative about the future, you begin to create the need for your customers to have those products of the future.

Critical Thinking Question

- What are ways that you might find out information about the products and services your company will be offering in the future?

Use of Effective Sales Strategies

Effective selling is more than just talking about a new product or service. There are several proven strategies that you can use to move the customer from not knowing about a product to being interested to having a need to making the actual purchase. Let's explore some variations of effective sales strategies.

Hierarchy of Effects Model

The **Hierarchy of Effects Model** for sales and advertising has been around for decades. This model for selling suggests that an individual goes through a series of thought stages in developing the decision to make a purchase. The steps in the Hierarchy of Effects are illustrated in the Figure 6–1. The steps can be easily remembered using the acronym **AIDA.**

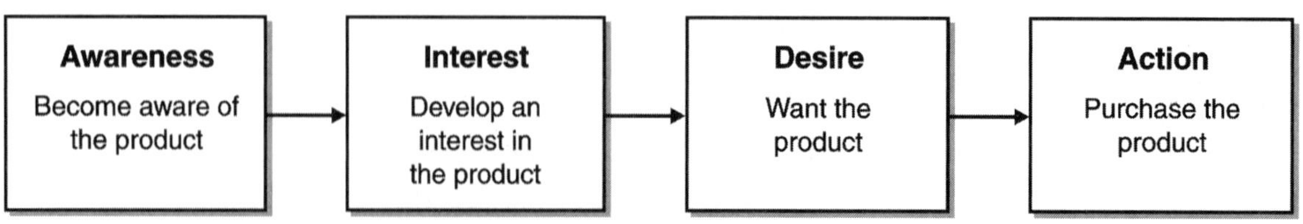

FIGURE 6–1 Hierarchy of Effects Model Diagram

These steps should be looked at with flexibility and in light of each selling opportunity. There are many variations to this model in the marketing and advertising industry; however, for simplicity, it is important to simply understand the basic steps.

- **Awareness:** Before a customer can make a decision to purchase something, they must be aware that it is available for purchase. This first step is especially important in the constantly changing area of technology. It is difficult for technology professionals to keep up with the new products in the industry. Customers typically do not spend the time it takes to stay abreast of what is available. Obviously, to develop awareness, you must first be aware of the products yourself and be able to explain them to your customers in terms they can understand.

  To make customers aware of new opportunities, you must develop a relationship with them so they will listen to what you have to say. You must also observe and communicate with your customers so that you can get an idea of the products and services that might be appropriate.
- **Interest:** Just because individuals are aware of something does not mean that they are interested in it. To move from awareness to decision to purchase, they must become interested in the product and begin to relate the product to their own unique situation. To develop interest, you must encourage customers to start asking questions about how it works, how it is useful, how much it costs, how easy it is to use, how it might meet their needs, and so forth.
- **Desire:** Customers must then desire the product in order to finally decide to make the purchase. Interest does not necessarily equate with desire. The answers to all the questions must lead to really wanting or thinking they need the product before they make the decision to purchase it. To help customers develop this desire, you must be enthusiastic about the products and help them see the benefits by the information you provide or even a demonstration of the products.
- **Action:** Finally, the customers must take some action that will eventually lead to purchase of the product. Action steps might include asking for a brochure, going to a Web site for more information, talking to sales people about the product, viewing a demonstration, talking to family members, and so forth. If some kind of action does not take place, the process toward purchase is stopped. You can help your customers take the action steps by suggesting an appropriate action. To do this, you must know what actions are available and which ones have a better chance of leading to a decision to purchase. Consider talking to your company's sales staff.

Features, Advantages, and Benefits (FAB)

Another useful way to think about selling additional products and services is in the 1960's **FAB (Features, Advantages, and Benefits)** sales concept. In this model, you help the customer become aware of the different features of a new product and how these features are advantages

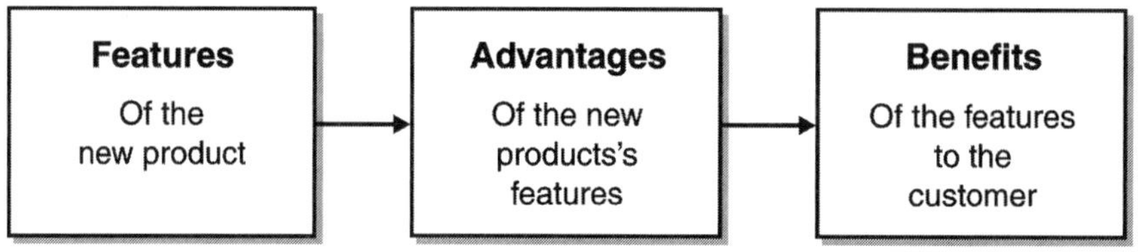

FIGURE 6–2 FAB Diagram

over other similar products and in general. Then the customer needs to see how the advantages are beneficial in meeting the customer's needs. Figure 6–2 illustrates the basic concept of the FAB model.

- **Features:** First, the product should be explained in terms of the various features. A **feature** is an observable characteristic. Exploring features is a good way to compare different products offered by your company and your products to those of the competition. Exploring features is also an excellent way to educate the customer and plant seeds for the future. For example, a customer may have a TV and your company may sell a plasma TV that is compatible with your computer.
- **Advantages:** The next step is to explain the advantages of the features of the product. An **advantage** is what the feature does for the purchaser. In our plasma TV example, the feature of computer compatibility offers the advantage of turning the plasma TV into a computer monitor.
- **Benefits:** Finally, frame the advantages in terms of benefits to the customer. A **benefit** is the payoff for the purchases. How does the product make their life easier, save them money, free up space, allow specific activities, and so forth. In our example, the advantage of turning the plasma TV into a computer monitor is that you can now enjoy inexpensive movies downloaded from the Internet, you can view slide shows of family photos directly from your computer, you can play multiple player computer games more easily, and so forth. The features of the plasma TV can be combined with wireless technologies resulting in additional sales.

SPIN® Method of Selling

Neil Rackham conducted an extensive study on the best selling strategies used by successful sales staff and developed the **SPIN®** strategy of selling. Figure 6–3 illustrates this simple and logical method. SPIN stands for Situation, Problem, Implication, Need or Need Payoff.

Unique Selling Point (USP)

A **USP** or unique selling point is the difference your product has from other similar products. As suggested, it separates your product from others. Continuing with the plasma TV example, if very few TVs were able to connect to a computer, then this feature would be a USP. It is important to note that focusing on the unique selling point is only a good selling strategy if the customer cares about that feature. Also, price is not a good USP, especially in technology areas. Typically, consumers do not purchase the lowest priced product. The features are

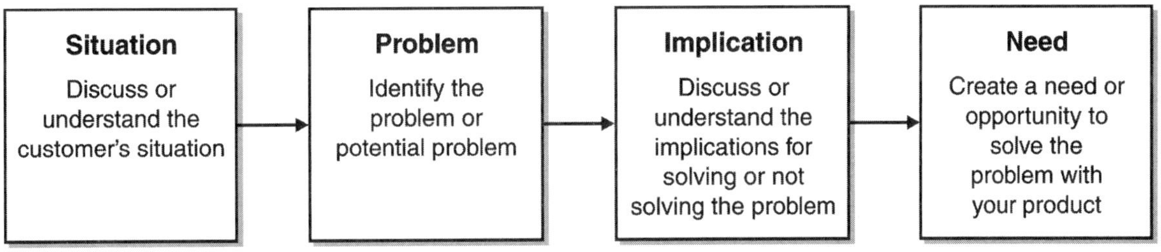

FIGURE 6–3 SPIN® Diagram *(Acquired with permission. Information about SPIN® is used with Huthwaite's permission, and shall not be reproduced or distributed. Huthwaite®, SPIN®, and SPIN Selling® are trademarks owned by Huthwaite and used with Huthwaite's permission. If you would like further information about either Huthwaite or its offerings, they can be contacted at Huthwaite, Inc.,* ***www.huthwaite.com, info@huthwaite.com.****)*

much more important. Additional examples of unique selling points include the inclusion of training, free installation, or extended warranties with the purchase.

Unique Perceived Benefit (UPB)

The **UPB** or unique perceived benefit is a selling strategy that focuses on the actual or perceived needs of the customer. The UPB may be different for different customers. In the plasma TV example, let's assume that your customer has a large family with several children and the grandparents visit often. The family also uses their digital camera all the time. For this family, a unique perceived benefit is that they can easily project their digital photos directly from their computer to the plasma TV for easy viewing by the family and grandparents. Since they love to show slide shows, this TV is perfect for the family. Another unique perceived benefit might be that the parents can more easily monitor what the kids are doing on the Internet if the computer screen is projected to the plasma TV in the living room.

Your Company's Sales Procedures

If sales is part of your job description, you will know the procedures for selling products and services to your customers. If selling is not a part of your job description, you should nonetheless understand the appropriate procedures so that the sales opportunity with your customer does not fall through the cracks. Every company is different. Some companies actively encourage every employee to be a sales representative for the company. Other companies want only sales representatives to sell products and services. Find out the business structure of your company and their policies and procedures. Talk to your supervisor about how to go about ensuring that a potential sale with your customer is routed to the right people or department if the opportunity occurs. It may be that you should call a sales representative and provide the appropriate information. If so, make friends with the sales staff so you can easily contact them when a sales opportunity occurs. They will love you for it. It may be that you can handle the sales directly by completing the appropriate paperwork or taking care of the transaction yourself over the phone or Internet.

It is important to follow your company's procedures carefully. Policies and procedures are generally established for a reason. There are many common problems that can occur when employees do not follow the proper channels.

- The employee might provide incorrect product information, an outdated price or service fee, or information about a product that has been discontinued.
- The employee might infringe on a sale that legitimately belongs to a sales representative causing hard feelings or even more serious consequences.
- The employee might not be aware of a discount, rebate, or free service for which the customer is qualified.
- The employee might complete paperwork incorrectly causing problems for the customer such as lost documentation, incorrect ordering information, privacy violations, delayed deliveries, and so forth.

Closing a Sale

After you have developed a rapport with your customer, determined potential existing or future needs or interests, explained appropriate products in terms of advantages and benefits, and listened to the customer's thoughts, there is an opportunity to close a sale. Many customers do not purchase a product or service because they were never asked if they wanted to. Asking the customer if they want to make the purchase is called **closing the sale.** Closing the sale is not as scary as it seems.

So, Mrs. Adams, would you like me to help you place an order for the upgrade? We can do it online right now, if you want.

If this new system looks like it will meet your needs, how about if I call the sales representative I work with, Mike Jackson, right now and we can get the order in. I can explain to him exactly what we discussed.

Going the Extra Mile

Going the extra mile means to give more than 100 percent and more than the customer is expecting. It is common for the rewards of this practice to be far more than the extra effort required. Let's explore some ideas and benefits associated with going the extra mile in your service to customers.

Critical Thinking Questions

Think back to a situation in which you got more than you expected, more than you paid for, or more than was promised.

- How did you feel?
- What was the result?
- Did you tell anyone else?
- What was their response?

Go the Extra Mile with Added Value

Added value is an element of a product or service that a company provides to help in the sale of the product or service. For example, free training on the use of a home entertainment center or free installation with purchase of the system is added value. An added value may be perceived or real; however, the customer must see it as a benefit. Added value is often used to attract new customers, take customers away from the competition, or keep existing customers. Added value comes in many forms. Research your competition to see what they offer or do not offer. Here are some examples:

- Make things easy (processes, procedures, paperwork, "red tape," ordering, scheduling, and so forth).
- Think about ways to make your tasks more efficient and then implement these ideas.
- Give out free information and resources.
- Assure your customers that they will get your personal attention. Follow up on services, questions, problems, complaints, and satisfaction promptly and on a regular basis.

Go the Extra Mile with Attention to Detail

Paying attention to the smallest details is an extra way to provide your customers with extra service. Put yourself in your customer's shoes for a minute. Look around. What details can you see? Here are some examples:

- If you are entering a customer's home, make sure you do not track in dirt or water or snow from outside. Put coverings on your shoes or lay down coverings on the floors.
- After you have completed an installation project, clean up after yourself. Consider running a handheld vacuum over your work area, wipe down counters, walls, and equipment with attention to smudges or fingerprints, and so forth.
- Put furniture back where you found it or offer to help the customer move furniture to the correct position.
- Set clocks, channels, or other programmable features like the customer wants.
- Leave the customer your card or contact information so they can easily call you with questions.
- Send customers short notes thanking them for their business and congratulating them on their purchase, their new home, construction completion, etc.
- Remember your customers' names as well as those of their spouses and children (and even dog, if it is a major presence in the home).
- When appropriate, offer to fix a problem or perform a task even if it is not what you are paid to do.
- Make it your responsibility to be the knowledge expert on the technology and products you use. Encourage your customers to call you if they have any questions at all, even if not directly related to the immediate purchase.

Benefits of Going the Extra Mile

The benefits of going the extra mile with your customers far outweigh the extra effort. Going the extra mile

- Shows your customers that you value them and care about their needs
- Helps to secure a loyal customer for the long term
- Increases the likelihood that your customers will tell others about the great service you gave them and encourage them to do their business with you as well
- Helps to offset any problems that occurred
- Creates a culture of positive attitude, reduces the poor attitudes of customers, and minimizes customer complaints, which in turn makes your job easier
- Creates a relationship where your customers trust you to meet their needs.

Study/Career Tip

Go to the Internet and use **www.google.com** to conduct a search using the key words "going the extra mile".

- What do different companies mean when they use this in their marketing?

List as many different "extras" as you can find.

Critical Thinking Questions

Make your own list.

- What are things you think you could do in your own situation that would reflect going the extra mile for your customers?

Write down as many ideas as you can and continue to add to the list as you move through your course work and career.

- What do you think will be the return on this investment?

Make a list of ideas for how you think going the extra mile with your customers will benefit you personally.

Summary

In this chapter we discussed several important ingredients for sustaining a satisfied and loyal customer over the long term. First, we reviewed the importance and goals of following up with customers after completing a job or solving a problem as well as following up periodically with existing customers to see how they are and to open doors for the next sale. We then explored strategies for making that next sale. Finally, we looked at the benefits of going the extra mile and many ideas for doing so.

Important points in this chapter are:

- It is important to follow up with customers to ensure that they are satisfied.
- Follow-up should occur after a job to ensure that everything works as it should and the customer does not have questions. Follow-up

should also occur after you have solved a problem or answered a question to make sure that the customer is satisfied.

- Follow-up should occur periodically with existing customers to make them feel valued and to open doors for additional sales.
- There are several strategies that should be followed for excellent follow-up with customers.
- An important goal of excellent service is to make sure that the customer looks to your company for any additional products or service they want to purchase.
- Every interaction with a customer is an opportunity for the next sale, and all employees are in some way responsible for creating sales opportunities.
- Every employee should understand the sales opportunity as well as basic sales strategies that secure or plant seeds for the next sale.
- Every employee should also understand the company's policies and procedures for the sales process to ensure that the sale is not lost.
- Closing a sale is important and means that you ask the customer if they want to make the purchase.
- Going the extra mile with customers means giving more than 100 percent and giving the customers more than what they expect.
- Going the extra mile typically reaps much greater rewards than the extra effort required.

Key Terms

Added value Added value is an element of a product or service that a company provides to help in the sale of the product or service.

Advantage An advantage is what a feature does for a purchaser.

AIDA AIDA is the acronym associated with the steps in the Hierarchy of Effects Model for selling. AIDA stands for awareness, interest, desire, and action.

Benefit The payoff for purchases.

Closing the sale Closing the sale is asking the customer if they want to make the purchase.

FAB FAB is a model for selling that highlights the features of a product, the advantages those products provide, and the benefits the advantage give to the customer.

Feature An observable characteristic.

Going the extra mile Going the extra mile means to give more than 100 percent and more than the customer is expecting.

Hierarchy of Effects Model The Hierarchy of Effects Model of selling suggests that an individual goes through a series of steps to make a decision to purchase. These steps include awareness, interest, desire, and action.

SPIN® selling SPIN® selling is a model for selling that encourages understanding of the situation, identification of the problem, understanding of the implications of the problem, and creation of a need or opportunity to solve the problem with the product.

UPB The UPB or unique perceived benefit is a selling strategy that focuses on the actual or perceived needs of the customer.

USP A USP or unique selling point is the difference your product has from other similar products.

Review Questions

1. List nine service goals met by excellent follow-up with customers.
2. Explain eight follow-up strategies that can be used to ensure excellent service.
3. Identify five elements you should thoroughly understand to optimize any sales opportunity.
4. Why is it important to understand your customer's existing needs and interests?

5. Why is it important to understand your customer's future needs and interests?
6. What are specific ways to learn more about your company's products and services?
7. Outline the Hierarchy of Effects model for selling.
8. What is the FAB sales concept?
9. Outline the steps in the SPIN® method of selling.
10. What is the difference between USP and UPB?
11. Why is it important to understand your company's policies and procedures for selling?
12. Why is closing a sale important in the selling process?
13. Why would you want to go the extra mile with a customer who has already purchased your product?

Glossary

Added Value Added value is an element of a product or service that a company provides to help in the sale of the product or service.

Advantage An advantage is what a feature does for a purchaser.

AIDA AIDA is the acronym associated with the steps in the Hierarchy of Effects Model for selling. AIDA stands for awareness, interest, desire, and action.

Auditory learners Auditory learners are those who learn mainly by hearing.

Benefit The payoff for services.

Body language Body language is a set of cues that you give to others in the process of communication.

Business strategy A business strategy is a company's specific plan for doing business.

Chunking Chunking means to organize information into small groups and to make the items in the small groups relate to each other in some logical way.

Clarifying question A clarifying question is one that asks for more facts or description.

Closed-ended question A closed-ended question is one that asks for a definite answer such as yes or no or some other factual response.

Closing the Sale Closing the sale is asking the customer if they want to make the purchase.

Cognitive Apprenticeship Model The Cognitive Apprenticeship Model is an effective step-by-step approach to teaching hands-on skills.

Communication channel The way in which a communication is sent or the type of media that is used such as face-to-face, computer, e-mail, written document, telephone, etc.

Communication process The communication process outlines four steps that should be followed for effective communication. The process includes (1) encoding and sending of the message, (2) transmission of the message through an appropriate channel, (3) receiving and decoding of the message by the receiver, and (4) interpretation and understanding of the message by the receiver.

Customer service Customer service is the total effect of the products, services, and information you provide your customer combined with how you treat the customer before, during, and after the sale.

Empathy Empathy is putting yourself in the other person's shoes.

Evaluation question An evaluation question is one that can be used when you need your customer to offer their expert opinion.

Example question An example question is one that can be used when you do not understand the customer's explanation.

Excellent service The total effect of the products, services, and information you provide to your customer combined with how you treat the customer before, during, and after the sale or service.

Existing customer An existing customer is one who has purchased your products and services and who resides on your customer list.

Exiting customer An exiting customer is one who leaves for a definite reason, such as dissatisfaction, or just in the natural course of life.

Extension question An extension question is one that can be used when the customer has not provided sufficient information and you need more.

External customer One who is directly or indirectly pays for your products and services.

FAB FAB is a model for selling that highlights the features of a product, the advantages those products provide, and the benefits the advantage give to the customer.

Feature An observable characteristic.

Going the extra mile Going the extra mile means to give more than 100 percent and more than the customer is expecting.

Hierarchy of Effects Model The Hierarchy of Effects Model of selling suggests that an individual goes through a series of steps to make a decision to purchase. These steps include awareness, interest, desire, and action.

Internal customer An internal customer can be an individual, a department, or a partner who needs you something from you and on whom you rely to serve external customers successfully.

Kinesthetic learners Kinesthetic learners are those who learn mainly be using or doing.

Learning styles A learning style describes the way in which an individual learns best in most circumstances.

Life-long learning Life-long learning is the idea that a professional must continue developing knowledge and skills throughout their life in order to maintain competency for the workplace.

Mission statement A mission statement is a concise statement declaring the kind of business you are in and identifying the customers you serve. It is sometimes called a company vision.

Navigational learners Navigational learners understand the concept and reason behind each procedure and can reapply a set of procedures, perhaps in a different way, to perform a different function as the situation requires.

Netiquette Netiquette represents the manners or professional courtesies followed in an on-line communication.

New customer A new customer is one who has recently purchased a product or service.

Open-ended question An open-ended question is one that seeks more than a one or two word response.

Organizational chart An organizational chart is a diagram that lists all departments and positions within a company and illustrates how these positions interrelate and communicate with each other.

Organizational structure An organizational company structure describes a company's various tasks and the positions within a company and how these positions interrelate with each other.

Preview-Teach-Review Model The Preview-Teach-Review Model is very basic teaching strategy that encourages a trainer to follow three simple steps: (1) preview the information, (2) teach the information, and (3) review the information.

Probing question A probing question is one that solicits more information from the customer within a specific line of questioning or on a specific topic.

Problem solving Problem solving is a general term used to describe looking at a situation, determining the cause of the problem, developing potential solutions, and implementing the solution.

Problem Solving Process The problem solving process is a step-by-step system for ensuring that problems are solved in a logical, thorough, and efficient manner.

Procedural learners Procedural learners memorize a series of steps to achieve a specific function and apply these steps exactly any time they want to perform the function.

Project team A project team is any group of individuals who interact to achieve a specific purpose.

Purpose question A purpose question is one that can be used when you need to understand the underlying reasons for something.

Self-talk Self-talk is what you say silently to yourself as you go through the day or are faced with difficult situations.

SPIN® selling SPIN® selling is a model for selling that encourages understanding of the situation, identification of the problem, understanding of the implications of the problem, and creation of a need or opportunity to solve the problem with the product.

Troubleshooting Troubleshooting is often used for the problem solving activities in the technical aspects of a job, such as determining how a piece of equipment is malfunctioning.

UPB The UPB or unique perceived benefit is a selling strategy that focuses on the actual or perceived needs of the customer.

USP A USP or unique selling point is the difference your product has from other similar products.

Visual learners Visual learners are those who learn mainly by seeing diagrams and images.

Index